Nikolaos Karagiannis

Evolutionary Game Theory

We developed the basic ideas of game theory, in which individual players

make decisions, and the payoff to each player depends on the decisions made by all. As we

saw there, a key question in game theory is to reason about the behavior we should expect

to see when players take part in a given game.

The discussion in Chapter 6 was based on considering how players simultaneously reason

about what the other players may do. In this chapter, on the other hand, we explore the

notion of *evolutionary game theory*, which shows that the basic ideas of game theory can be

applied even to situations in which no individual is overtly reasoning, or even making explicit

decisions. Rather, game-theoretic analysis will be applied to settings in which individuals can exhibit different forms of behavior (including those that may not be the result of conscious

choices), and we will consider which forms of behavior have the ability to persist in the

population, and which forms of behavior have a tendency to be driven out by others.

As its name suggests, this approach has been applied most widely in the area of evolutionary

biology, the domain in which the idea was first articulated by John Maynard Smith

and G. R. Price [375, 376]. Evolutionary biology is based on the idea that an organism's

genes largely determine its observable characteristics, and hence its *fitness* in a given environment.

Organisms that are more fit will tend to produce more offspring, causing genes

that provide greater fitness to increase their representation in the population. In this way,

fitter genes tend to win over time, because they provide higher rates of reproduction.

The key insight of evolutionary game theory is that many behaviors involve the *interaction*

of multiple organisms in a population, and the success of any one of these organisms depends

on how its behavior interacts with that of others. So the fitness of an individual organism

can't be measured in isolation; rather it has to be evaluated in the context of the full

population in which it lives. This opens the door to a natural game-theoretic analogy:

Draft version: June 10, 2010

209

210 *CHAPTER 7. EVOLUTIONARY GAME THEORY*

an organism's genetically-determined characteristics and behaviors are like its strategy in a

game, its fitness is like its payoff, and this payoff depends on the strategies (characteristics) of

the organisms with which it interacts. Written this way, it is hard to tell in advance whether

this will turn out to be a superficial analogy or a deep one, but in fact the connections turn out to run very deeply: game-theoretic ideas like equilibrium will prove to be a useful way

to make predictions about the results of evolution on a population.

7.1 Fitness as a Result of Interaction

To make this concrete, we now describe a first simple example of how game-theoretic ideas

can be applied in evolutionary settings. This example will be designed for ease of explanation

rather than perfect fidelity to the underlying biology; but after this we will discuss examples

where the phenomenon at the heart of the example has been empirically observed in a variety

of natural settings.

For the example, let's consider a particular species of beetle, and suppose that each

beetle's fitness in a given environment is determined largely by the extent to which it can

find food and use the nutrients from the food effectively. Now, suppose a particular mutation

is introduced into the population, causing beetles with the mutation to grow a significantly larger body size. Thus, we now have two distinct kinds of beetles in the population — small

ones and large ones. It is actually difficult for the large beetles to maintain the metabolic

requirements of their larger body size — it requires diverting more nutrients from the food

they eat — and so this has a negative effect on fitness.

If this were the full story, we'd conclude that the large-body-size mutation is fitnessdecreasing,

and so it will likely be driven out of the population over time, through multiple

generations. But in fact, there's more to the story, as we'll now see.

Interaction Among Organisms. The beetles in this population compete with each other

for food – when they come upon a food source, there's crowding among the beetles as they

each try to get as much of the food as they can. And, not surprisingly, the beetles with large

body sizes are more effective at claiming an above-average share of the food.

Let's assume for simplicity that food competition in this population involves two beetles

interacting with each other at any given point in time. (This will make the ideas easier

to describe, but the principles we develop can also be applied to interactions among many

individuals simultaneously.) When two beetles compete for some food, we have the following

possible outcomes.

• When beetles of the same size compete, they get equal shares of the food.

• When a large beetle competes with a small beetle, the large beetle gets the majority

of the food.

7.2. EVOLUTIONARILY STABLE STRATEGIES 211

• In all cases, large beetles experience less of a fitness benefit from a given quantity of

food, since some of it is diverted into maintaining their expensive metabolism.

Thus, the fitness that each beetle gets from a given food-related interaction can be thought of as a numerical payoff in a two-player game between a first beetle and a second

beetle, as follows. The first beetle plays one of the two strategies *Small* or *Large*, depending

on its body size, and the second beetle plays one of these two strategies as well. Based on

the two strategies used, the payoffs to the beetles are described by Figure 7.1.

Beetle 1

Beetle 2

Small Large

Small 5, 5 1, 8

Large 8, 1 3, 3

Figure 7.1: The Body-Size Game

Notice how the numerical payoffs satisfy the principles just outlined: when two small

beetles meet, they share the fitness from the food source equally; large beetles do well at

the expense of small beetles; but large beetles cannot extract the full amount of fitness from the food source. (In this payoff matrix, the reduced fitness when two large beetles meet is

particularly pronounced, since a large beetle has to expend extra energy in competing with

another large beetle.)

This payoff matrix is a nice way to summarize what happens when two beetles meet,

but compared with the game in Chapter 6, there's something fundamentally different in

what's being described here. The beetles in this game aren't asking themselves, "What do

I want my body size to be in this interaction?" Rather, each is genetically hard-wired to

play one of these two strategies through its whole lifetime. Given this important difference,

the idea of choosing strategies — which was central to our formulation of game theory —

is missing from the biological side of the analogy. As a result, in place of the idea of Nash

equilibrium — which was based fundamentally on the relative benefit of changing one's own personal strategy — we will need to think about strategy changes that operate over longer

time scales, taking place as shifts in a population under evolutionary forces. We develop the

fundamental definitions for this in the next section.

7.2 Evolutionarily Stable Strategies

In Chapter 6, the notion of Nash equilibrium was central in reasoning about the outcome

of a game. In a Nash equilibrium for a two-player game, neither player has an incentive to

deviate from the strategy they are currently using — the equilibrium is a choice of strategies

that tends to persist once the players are using it. The analogous notion for evolutionary

212 CHAPTER 7. EVOLUTIONARY GAME THEORY

settings will be that of an *evolutionarily stable strategy* — a genetically-determined strategy

that tends to persist once it is prevalent in a population.

We formulate this as follows. Suppose, in our example, that each beetles is repeatedly paired off with other beetles in food competitions over the course of its lifetime. We will

assume the population is large enough that no two particular beetles have a significant

probability of interacting with each other repeatedly. A beetle's overall fitness will be equal

to the average fitness it experiences from each of its many pairwise interactions with others,

and this overall fitness determines its reproductive success — the number of offspring that

carry its genes (and hence its strategy) into the next generation.

In this setting, we say that a given strategy is *evolutionarily stable* if, when the whole

population is using this strategy, any small group of invaders using a different strategy

will eventually die off over multiple generations. (We can think of these invaders either

as migrants who move to join the population, or as mutants who were born with the new

behavior directly into the population.) We capture this idea in terms of numerical payoffs by saying that when the whole population is using a strategy S, then a small group of

invaders using any alternate strategy T should have strictly lower fitness than the users

of the majority strategy S. Since fitness translates into reproductive success, evolutionary

principles posit that strictly lower fitness is the condition that causes a sub-population (like

the users of strategy T) to shrink over time, through multiple generations, and eventually

die off with high probability.

More formally, we will phrase the basic definitions as follows.

• We say the *fitness* of an organism in a population is the expected payoff it receives

from an interaction with a random member of the population.

• We say that a strategy T *invades* a strategy S at level x, for some small positive

number x, if an x fraction of the underlying population uses T and a $1 - x$ fraction of

the underlying population uses S.

• Finally, we say that a strategy S is *evolutionarily stable* if there is a (small) positive

number y such that when any other strategy T invades S at any level $x < y$, the fitness

of an organism playing S is strictly greater than the fitness of an organism playing T.

Evolutionarily Stable Strategies in our First Example. Let's see what happens when

we apply this definition to our example involving beetles competing for food. We will first

check whether the strategy *Small* is evolutionarily stable, and then we will do the same for

the strategy *Large*.

Following the definition, let's suppose that for some small positive number x, a $1 - x$

fraction of the population uses *Small* and an x fraction of the population uses *Large*. (This

7.2. EVOLUTIONARILY STABLE STRATEGIES 213

is what the picture would look like just after a small invader population of large beetles

arrives.)

• What is the expected payoff to a small beetle in a random interaction in this population?

With probability $1 - x$, it meets another small beetle, receiving a payoff of 5,

while with probability x, it meets a large beetle, receiving a payoff of 1. Therefore its

expected payoff is

$5(1 - x) + 1 \cdot x = 5 - 4x.$

• What is the expected payoff to a large beetle in a random interaction in this population?

With probability $1 - x$, it meets a small beetle, receiving a payoff of 8, while with

probability x, it meets another large beetle, receiving a payoff of 3. Therefore its

expected payoff is

$8(1 - x) + 3 \cdot x = 8 - 5x.$

It's easy to check that for small enough values of x (and even for reasonably large ones

in this case), the expected fitness of large beetles in this population exceeds the expected fitness of small beetles. Therefore *Small* is not evolutionarily stable.

Now let's check whether *Large* is evolutionarily stable. For this, we suppose that for some

very small positive number x, a $1-x$ fraction of the population uses *Large* and an x fraction

of the population uses *Small*.

• What is the expected payoff to a large beetle in a random interaction in this population?

With probability $1-x$, it meets another large beetle, receiving a payoff of 3, while with

probability x, it meets a small beetle, receiving a payoff of 8. Therefore its expected

payoff is

$$3(1 - x) + 8 \cdot x = 3 + 5x.$$

• What is the expected payoff to a small beetle in a random interaction in this population?

With probability $1 - x$, it meets a large beetle, receiving a payoff of 1, while

with probability x, it meets another small beetle, receiving a payoff of 5. Therefore its expected payoff is

$$(1 - x) + 5 \cdot x = 1 + 4x.$$

In this case, the expected fitness of large beetles in this population exceeds the expected

fitness of small beetles, and so *Large* is evolutionarily stable.

214 *CHAPTER 7. EVOLUTIONARY GAME THEORY*

Interpreting the Evolutionarily Stable Strategy in our Example. Intuitively, this

analysis can be summarized by saying that if a few large beetles are introduced into a

population consisting of small beetles, then the large beetles do extremely well — since

they rarely meet each other, they get most of the food in almost every competition they

experience. As a result, the population of small beetles cannot drive out the large ones, and

so *Small* is not evolutionarily stable.

On the other hand, in a population of large beetles, a few small beetles will do very badly, losing almost every competition for food. As a result, the population of large beetles resists

the invasion of small beetles, and so *Large* is evolutionarily stable.

Therefore, if we know that the large-body-size mutation is possible, we should expect to

see populations of large beetles in the wild, rather than populations of small ones. In this

way, our notion of evolutionary stability has predicted a strategy for the population — as we

predicted outcomes for games among rational players in Chapter 6, but by different means.

What's striking about this particular predicted outcome, though, is the fact that the

fitness of each organism in a population of small beetles is 5, which is larger than the fitness

of each organism in a population of large beetles. In fact, the game between small and large

beetles has precisely the structure of a Prisoner's Dilemma game; the motivating scenario

based on competition for food makes it clear that the beetles are engaged in an arms race, like the game from Chapter 6 in which two competing athletes need to decide whether to use

performance-enhancing drugs. There it was a dominant strategy to use drugs, even though

both athletes understand that they are better off in an outcome where neither of them uses

drugs — it's simply that this mutually better joint outcome is not sustainable. In the present

case, the beetles individually don't understand anything, nor could they change their body

sizes even if they wanted to. Nevertheless, evolutionary forces over multiple generations are

achieving a completely analogous effect, as the large beetles benefit at the expense of the

small ones. Later in this chapter, we will see that this similarity in the conclusions of two

different styles of analysis is in fact part of a broader principle.

Here is a different way to summarize the striking feature of our example: Starting from

a population of small beetles, evolution by natural selection is causing the fitness of the organisms to decrease over time. This might seem troubling initially, since we think of

natural selection as being fitness-increasing. But in fact, it's not hard to reconcile what's

happening with this general principle of natural selection. Natural selection increases the

fitness of individual organisms in a fixed environment — if the environment changes to

become more hostile to the organisms, then clearly this could cause their fitness to go down.

This is what is happening to the population of beetles. Each beetle's environment includes

all the other beetles, since these other beetles determine its success in food competitions;

therefore the increasing fraction of large beetles can be viewed, in a sense, as a shift to an

environment that is more hostile for everyone.

7.2. EVOLUTIONARILY STABLE STRATEGIES 215

Empirical Evidence for Evolutionary Arms Races. Biologists have offered recent evidence

for the presence of evolutionary games in nature with the Prisoner's-Dilemma structure we've just seen. It is very difficult to truly determine payoffs in any real-world setting, and so

all of these studies are the subject of ongoing investigation and debate. For our purposes in

this discussion, they are perhaps most usefully phrased as deliberately streamlined examples,

illustrating how game-theoretic reasoning can help provide qualitative insight into different

forms of biological interaction.

It has been argued that the heights of trees can obey Prisoner's-Dilemma payoffs [156,

226]. If two neighboring trees both grow short, then they share the sunlight equally. They

also share the sunlight equally if they both grow tall, but in this case their payoffs are each

lower because they have to invest a lot of resources in achieving the additional height. The

trouble is that if one tree is short while its neighbor is tall, then the tall tree gets most of

the sunlight. As a result, we can easily end up with payoffs just like the Body-Size Game among beetles, with the trees' evolutionary strategies *Short* and *Tall* serving as analogues to

the beetles' strategies *Small* and *Large*. Of course, the real situation is more complex than

this, since genetic variation among trees can lead to a wide range of different heights and

hence a range of different strategies (rather than just two strategies labeled *Short* and *Tall*).

Within this continuum, Prisoner's-Dilemma payoffs can only apply to a certain range of tree

heights: there is some height beyond which further height-increasing mutations no longer

provide the same payoff structure, because the additional sunlight is more than offset by the

fitness downside of sustaining an enormous height.

Similar kinds of competition take place in the root systems of plants [181]. Suppose you

grow two soybean plants at opposite ends of a large pot of soil; then their root systems

will each fill out the available soil and intermingle with each other as they try to claim as many resources as they can. In doing so, they divide the resources in the soil equally. Now,

suppose that instead you partition the same quantity of soil using a wall down the middle, so

that the two plants are on opposite sides of the wall. Then each still gets half the resources

present in the soil, but each invests less of its energy in producing roots and consequently

has greater reproductive success through seed production.

This observation has implications for the following simplified evolutionary game involving

root systems. Imagine that instead of a wall, we had two kinds of root-development strategies

available to soybean plants: *Conserve*, where a plant's roots only grow into its own share of

the soil, and *Explore*, where the roots grow everywhere they can reach. Then we again have

the scenario and payoffs from the Body-Size Game, with the same conclusion: all plants are

better off in a population where everyone plays *Conserve*, but only *Explore* is evolutionarily stable.

As a third example, there was recent excitement over the discovery that virus populations

can also play an evolutionary version of the Prisoner's Dilemma [326, 392]. Turner and Chao

216 *CHAPTER 7. EVOLUTIONARY GAME THEORY*

studied a virus called Phage #6, which infects bacteria and manufactures products needed

for its own replication. A mutational variant of this virus called Phage #H2 is also able to

replicate in bacterial hosts, though less effectively on its own. However, #H2 is able to take

advantage of chemical products produced by #6, which gives #H2 a fitness advantage when

it is in the presence of #6. This turns out to yield the structure of the Prisoner's Dilemma:

viruses have the two evolutionary strategies #6 and #H2; viruses in a pure #6 population all

do better than viruses in a pure #H2 population; and regardless of what the other viruses are

doing, you (as a virus) are better off playing #H2. Thus only #H2 is evolutionarily stable.

The virus system under study was so simple that Turner and Chao were able to infer an

actual payoff matrix based on measuring the relative rates at which the two viral variants

were able to replicate under different conditions. Using an estimation procedure derived

from these measurements, they obtained the payoffs in Figure 7.2. The payoffs are re-scaled

so that the upper-left box has the value 1.00, 1.00.1

Virus 1

Virus 2

#6 #H2

#6 1.00, 1.00 0.65, 1.99

#H2 1.99, 0.65 0.83, 0.83

Figure 7.2: The Virus Game

Whereas our earlier examples had an underlying story very much like the use of performanceenhancing

drugs, this game among phages is actually reminiscent of a different story that also motivates the Prisoner's Dilemma payoff structure: the scenario behind the Exam-or-

Presentation game with which we began Chapter 6. There, two college students would both

be better off if they jointly prepared for a presentation, but the payoffs led them to each think

selfishly and study for an exam instead. What the Virus Game here shows is that shirking

a shared responsibility isn't just something that rational decision-makers do; evolutionary

forces can induce viruses to play this strategy as well.

7.3 A General Description of Evolutionarily Stable Strategies

The connections between evolutionary games and games played by rational participants are

suggestive enough that it makes sense to understand how the relationship works in general.

We will focus here, as we have thus far, on two-player two-strategy games. We will also

1It should be noted that even in a system this simple, there are many other biological factors at work, and hence this payoff matrix is still just an approximation to the performance of "6 and "H2 populations

under real experimental and natural conditions. Other factors appear to affect these populations, including

the density of the population and the potential presence of additional mutant forms of the virus [393].

7.3. A GENERAL DESCRIPTION OF EVOLUTIONARILY STABLE STRATEGIES217

restrict our attention to symmetric games, as in the previous sections of this chapter, where

the roles of the two players are interchangeable.

The payoff matrix for a completely general two-player, two-strategy game that is symmetric

can be written as in Figure 7.3.

Organism 1

Organism 2

S T

S a, a b, c

T c, b d, d

Figure 7.3: General Symmetric Game

Let's check how to write the condition that *S* is evolutionarily stable in terms of the four

variables *a*, *b*, *c*, and *d*. As before, we start by supposing that for some very small positive

number *x*, a 1–*x* fraction of the population uses *S* and an *x* fraction of the population uses

T.

• What is the expected payoff to an organism playing *S* in a random interaction in this

population? With probability 1–*x*, it meets another player of *S*, receiving a payoff of

a, while with probability *x*, it meets a player of *T*, receiving a payoff of *b*. Therefore

its expected payoff is

$a(1 - x) + bx.$

• What is the expected payoff to an organism playing *T* in a random interaction in this

population? With probability 1–*x*, it meets a player of *S*, receiving a payoff of *c*, while with probability x, it meets another player of T, receiving a payoff of d. Therefore its

expected payoff is

$c(1 - x) + dx.$

Therefore, S is evolutionarily stable if for all sufficiently small values of $x > 0$, the

inequality

$a(1 - x) + bx > c(1 - x) + dx$

holds. As x goes to 0, the left-hand side becomes a and the right-hand side becomes c.

Hence, if $a > c$, then the left-hand side is larger once x is sufficiently small, while if $a < c$

then the left-hand side is smaller once x is sufficiently small. Finally, if $a = c$, then the

left-hand side is larger precisely when $b > d$. Therefore, we have a simple way to express the

condition that S is evolutionarily stable:

In a two-player, two-strategy, symmetric game, S is evolutionarily stable precisely

when either (i) $a > c$, or (ii) $a = c$ and $b > d$.

218 *CHAPTER 7. EVOLUTIONARY GAME THEORY*

It is easy to see the intuition behind our calculations that translates into this condition,

as follows.

• First, in order for S to be evolutionarily stable, the payoff to using strategy S against

S must be at least as large as the payoff to using strategy T against S. Otherwise, an

invader who uses T would have a higher fitness than the rest of population, and the

fraction of the population who are invaders would have a good probability of growing

over time.

• Second, if S and T are equally good responses to S, then in order for S to be evolutionarily

stable, players of S must do better in their interactions with T than players

of T do with each other. Otherwise, players of T would do as well as against the S

part of the population as players of S, and at least as well against the T part of the population, so their overall fitness would be at least as good as the fitness of players

of S.

7.4 Relationship Between Evolutionary and Nash Equilibria

Using our general way of characterizing evolutionarily stable strategies, we can now understand

how they relate to Nash equilibria. If we go back to the General Symmetric Game

from the previous section, we can write down the condition for (S, S) (i.e. the choice of S

by both players) to be a Nash equilibrium. (S, S) is a Nash equilibrium when S is a best

response to the choice of S by the other player: this translates into the simple condition

$a " c.$

If we compare this to the condition for S to be evolutionarily stable,

(i) $a > c$, or (ii) $a = c$ and $b > d,$

we immediately get the conclusion that

If strategy S is evolutionarily stable, then (S, S) is a Nash equilibrium.

We can also see that the other direction does not hold: it is possible to have a game

where (S, S) is a Nash equilibrium, but S is not evolutionarily stable. The difference in the

two conditions above tells us how to construct such a game: we should have $a = c$ and $b < d$.

To get a sense for where such a game might come from, let's recall the Stag Hunt Game

from Chapter 6. Here, each player can hunt stag or hunt hare; hunting hare successfully just

requires your own effort, while hunting the more valuable stag requires that you both do so.

This produces payoffs as shown in Figure 7.4.

7.4. RELATIONSHIP BETWEEN EVOLUTIONARY AND NASH EQUILIBRIA 219

Hunter 1

Hunter 2

Hunt Stag Hunt Hare

Hunt Stag 4, 4 0, 3

Hunt Hare 3, 0 3, 3

Figure 7.4: Stag Hunt

In this game, as written, *Hunt Stag* and *Hunt Hare* are both evolutionarily stable, as we

can check from the conditions on *a, b, c,* and *d.* (To check the condition for *Hunt Hare*, we

simply need to interchange the rows and columns of the payoff matrix, to put *Hunt Hare* in

the first row and first column.)

However, suppose we make up a modification of the Stag Hunt Game, by shifting the

payoffs as follows. In this new version, when the players mis-coordinate, so that one hunts

stag while the other hunts hare, then the hare-hunter gets an extra benefit due to the lack

of competition for hare. In this way, we get a payoff matrix as in Figure 7.5.

Hunter 1

Hunter 2

Hunt Stag Hunt Hare

Hunt Stag 4, 4 0, 4

Hunt Hare 4, 0 3, 3

Figure 7.5: Stag Hunt: A version with added benefit from hunting hare alone

In this case, the choice of strategies *(Hunt Stag, Hunt Stag)* is still a Nash equilibrium:

if each player expects the other to hunt stag, then hunting stag is a best response. But

Hunt Stag is not an evolutionarily stable strategy for this version of the game, because (in

the notation from our General Symmetric Game) we have $a = c$ and $b < d$. Informally,

the problem is that a hare-hunter and a stag-hunter do equally well when each is paired

with a stag-hunter; but hare-hunters do better than stag-hunters when each is paired with

a hare-hunter.

There is also a relationship between evolutionarily stable strategies and the concept of a

strict Nash equilibrium. We say that a choice of strategies is a strict Nash equilibrium if each player is using the unique best response to what the other player is doing. We can check

that for symmetric two-player, two-strategy games, the condition for (S, S) to be a strict

Nash equilibrium is that $a > c$. So we see that in fact these different notions of equilibrium

naturally *refine* each other. The concept of an evolutionarily stable strategy can be viewed as

a refinement of the concept of a Nash equilibrium: the set of evolutionarily stable strategies

S is a subset of the set of strategies S for which (S, S) is a Nash equilibrium. Similarly, the

concept of a strict Nash equilibrium (when the players use the same strategy) is a refinement

of evolutionary stability: if (S, S) is a strict Nash equilibrium, then S is evolutionarily stable.

220 *CHAPTER 7. EVOLUTIONARY GAME THEORY*

It is intriguing that, despite the extremely close similarities between the conclusions

of evolutionary stability and Nash equilibrium, they are built on very different underlying stories. In a Nash equilibrium, we consider players choosing mutual best responses to each

other's strategy. This equilibrium concept places great demands on the ability of the players

to choose optimally and to coordinate on strategies that are best responses to each other.

Evolutionary stability, on the other hand, supposes no intelligence or coordination on the part

of the players. Instead, strategies are viewed as being hard-wired into the players, perhaps

because their behavior is encoded in their genes. According to this concept, strategies which

are more successful in producing offspring are selected for.

Although this evolutionary approach to analyzing games originated in biology, it can be

applied in many other contexts. For example, suppose a large group of people are being

matched repeatedly over time to play the General Symmetric Game from Figure 7.3. Now

the payoffs should be interpreted as reflecting the welfare of the players, and not their number of offspring. If any player can look back at how others have played and can observe

their payoffs, then imitation of the strategies that have been most successful may induce

an evolutionary dynamic. Alternatively, if a player can observe his own past successes and

failures then his learning may induce an evolutionary dynamic. In either case, strategies that

have done relatively well in the past will tend to be used by more people in the future. This

can lead to the same behavior that underlies the concept of evolutionarily stable strategies,

and hence can promote the play of such strategies.

7.5 Evolutionarily Stable Mixed Strategies

As a further step in developing an evolutionary theory of games, we now consider how to

handle cases in which no strategy is evolutionarily stable.

In fact, it is not hard to see how this can happen, even in two-player games that have

pure-strategy Nash equilibria.2 Perhaps the most natural example is the Hawk-Dove Game from Chapter 6, and we use this to introduce the basic ideas of this section. Recall that in

the Hawk-Dove Game, two animals compete for a piece of food; an animal that plays the

strategy *Hawk* (*H*) behaves aggressively, while an animal that plays the strategy *Dove* (*D*)

behaves passively. If one animal is aggressive while the other is passive, then the aggressive

animal benefits by getting most of the food; but if both animals are aggressive, then they

risk destroying the food and injuring each other. This leads to a payoff matrix as shown in

Figure 7.6.

In Chapter 6, we considered this game in contexts where the two players were making

choices about how to behave. Now let's consider the same game in a setting where each

2Recall that a player is using a *pure strategy* if she always plays a particular one of the strategies in the

game, as opposed to a *mixed strategy* in which she chooses at random from among several possible strategies.

7.5. EVOLUTIONARILY STABLE MIXED STRATEGIES 221

Animal 1

Animal 2

D H

D 3, 3 1, 5

H 5, 1 0, 0

Figure 7.6: Hawk-Dove Game

animal is genetically hard-wired to play a particular strategy. How does it look from this

perspective, when we consider evolutionary stability?

Neither D nor H is a best response to itself, and so using the general principles from

the last two sections, we see that neither is evolutionarily stable. Intuitively, a hawk will do

very well in a population consisting of doves — but in a population of all hawks, a dove will

actually do better by staying out of the way while the hawks fight with each other.

As a two-player game in which players are actually choosing strategies, the Hawk-Dove

Game has two pure Nash equilibria: (D,H) and (H,D). But this doesn't directly help us

identify an evolutionarily stable strategy, since thus far our definition of evolutionary stability

has been restricted to populations in which (almost) all members play the same pure strategy.

To reason about what will happen in the Hawk-Dove Game under evolutionary forces, we

need to generalize the notion of evolutionary stability by allowing some notion of "mixing"

between strategies.

Defining Mixed Strategies in Evolutionary Game Theory. There are at least two

natural ways to introduce the idea of mixing into the evolutionary framework. First, it

could be that each individual is hard-wired to play a pure strategy, but some portion of

the population plays one strategy while the rest of the population plays another. If the fitness of individuals in each part of the population is the same, and if invaders eventually

die off, then this could be considered to exhibit a kind of evolutionary stability. Second,

it could be that each individual is hard-wired to play a particular mixed strategy — that

is, they are genetically configured to choose randomly from among certain options with

certain probabilities. If invaders using any other mixed strategy eventually die off, then this

too could be considered a kind of evolutionary stability. We will see that for our purposes

here, these two concepts are actually equivalent to each other, and we will focus initially

on the second idea, in which individuals use mixed strategies. Essentially, we will find that

in situations like the Hawk-Dove game, the individuals or the population as a whole must

display a mixture of the two behaviors in order to have any chance of being stable against

invasion by other forms of behavior.

The definition of an evolutionarily stable mixed strategy is in fact completely parallel

to the definition of evolutionary stability we have seen thus far — it is simply that we now

greatly enlarge the set of possible strategies, so that each strategy corresponds to a particular

222 CHAPTER 7. EVOLUTIONARY GAME THEORY

randomized choice over pure strategies.

Specifically, let's consider the General Symmetric Game from Figure 7.3. A mixed strategy

here corresponds to a probability p between 0 and 1, indicating that the organism plays S

with probability p and plays T with probability $1-p$. As in our discussion of mixed strategies

from Chapter 6, this includes the possibility of playing the pure strategies S or T by simply

setting $p = 1$ or $p = 0$. When Organism 1 uses the mixed strategy p and Organism 2 uses

the mixed strategy q, the expected payoff to Organism 1 can be computed as follows. There

is a probability pq of an (X,X) pairing, yielding a for the first player; there is a probability

$p(1-q)$ of an (X, Y) pairing, yielding b for the first player; there is a probability $(1-p)q$ of

a (Y,X) pairing, yielding c for the first player; and there is a probability $(1 - p)(1 - q)$ of a

(Y, Y) pairing, yielding d for the first player. So the expected payoff for this first player is

$$V(p, q) = pqa + p(1 - q)b + (1 - p)qc + (1 - p)(1 - q)d.$$

As before, the *fitness* of an organism is its expected payoff in an interaction with a random

member of the population. We can now give the precise definition of an evolutionarily stable

mixed strategy.

In the General Symmetric Game, p is an evolutionarily stable mixed strategy if

there is a (small) positive number y such that when any other mixed strategy q

invades p at any level x < y, the fitness of an organism playing p is strictly

greater than the fitness of an organism playing q.

This is just like our previous definition of evolutionarily stable (pure) strategies, except that we allow the strategy to be mixed, *and* we allow the invaders to use a mixed strategy.

An evolutionarily stable mixed strategy with $p = 1$ or $p = 0$ is evolutionarily stable under

our original definition for pure strategies as well. However, note the subtle point that even

if S were an evolutionarily stable strategy under our previous definition, it is not necessarily

an evolutionarily stable mixed strategy under this new definition with $p = 1$. The problem

is that it is possible to construct games in which no pure strategy can successfully invade a

population playing S, but a mixed strategy can. As a result, it will be important to be clear

in any discussion of evolutionary stability on what kinds of behavior an invader can employ.

Directly from the definition, we can write the condition for p to be an evolutionarily

stable mixed strategy as follows: for some y and any $x < y$, the following inequality holds

for all mixed strategies $q \#= p$:

$(1 - x)V(p, p) + xV(p, q) > (1 - x)V(q, p) + xV(q, q).$
(7.1)

This inequality also makes it clear that there is a relationship between mixed Nash

equilibria and evolutionarily stable mixed strategies, and this relationship parallels the one

we saw earlier for pure strategies. In particular, if p is an evolutionarily stable mixed strategy,

7.5. EVOLUTIONARILY STABLE MIXED STRATEGIES 223

then we must have $V(p, p)$ " $V(q, p)$, and so p is a best response to p. As a result, the pair

of strategies (p, p) is a mixed Nash equilibrium. However, because of the strict inequality

in Equation (7.1), it is possible for (p, p) to be a mixed Nash equilibrium without p being

evolutionarily stable. So again, evolutionary stability serves as a refinement of the concept

of mixed Nash equilibrium.

Evolutionarily Stable Mixed Strategies in the Hawk-Dove Game. Now let's see how to apply these ideas to the Hawk-Dove Game. First, since any evolutionarily stable

mixed strategy must correspond to a mixed Nash equilibrium of the game, this gives us a

way to search for possible evolutionarily stable strategies: we first work out the mixed Nash

equilibria for the Hawk-Dove, and then we check if they are evolutionarily stable.

As we saw in Chapter 6, in order for (p, p) to be a mixed Nash equilibrium, it must make

the two players indifferent between their two pure strategies. When the other player is using

the strategy p, the expected payoff from playing D is $3p+(1-p) = 1+2p$, while the expected

payoff from playing H is $5p$. Setting these two quantities equal (to capture the indifference

between the two strategies), we get $p = 1/3$. So $(1/3, 1/3)$ is a mixed Nash equilibrium. In

this case, both pure strategies, as well as any mixture between them, produce an expected

payoff of 5/3 when played against the strategy $p = 1/3$.

Now, to see whether $p = 1/3$ is evolutionarily stable, we must check Inequality (7.1)

when some other mixed strategy q invades at a small level x. Here is a first observation

that makes evaluating this inequality a bit easier. Since (p, p) is a mixed Nash equilibrium

that uses both pure strategies, we have just argued that all mixed strategies q have the same

payoff when played against p. As a result, we have $V(p, p) = V(q, p)$ for all q. Subtracting

these terms from the left and right of Inequality (7.1), and then dividing by x, we get the

following inequality to check:

$V(p, q) > V(q, q)$. (7.2)

The point is that since (p, p) is a mixed equilibrium, the strategy p can't be a strict best

response to itself — all other mixed strategies are just as good against it. Therefore, in order

for p to be evolutionarily stable, it must be a strictly better response to every other mixed

strategy q than q is to itself. That is what will cause it to have higher fitness when q invades.

In fact, it is true that $V(p, q) > V(q, q)$ for all mixed strategies $q \neq p$, and we can check

this as follows. Using the fact that $p = 1/3$, we have

$$V(p, q) = 1/3 \cdot q \cdot 3 + 1/3(1 - q) \cdot 1 + 2/3 \cdot q \cdot 5 = 4q + 1/3$$

while

$$V(q, q) = q2 \cdot 3 + q(1 - q) \cdot 1 + (1 - q) \cdot q \cdot 5 = 6q - 3q2.$$

Now we have

$$V(p, q) - V(q, q) = 3q2 - 2q + 1/3 =$$

1

3

$$(9q2 - 6q + 1) =$$

1

3

$$(3q - 1)2.$$

224 *CHAPTER 7. EVOLUTIONARY GAME THEORY*

This last way of writing $V(p, q) - V(q, q)$ shows that it is a perfect square, and so it is positive whenever $q \neq 1/3$. This is just what we want for showing that $V(p, q) > V(q, q)$ whenever

$q \neq p$, and so it follows that p is indeed an evolutionarily stable mixed strategy.

Interpretations of Evolutionarily Stable Mixed Strategies. The kind of mixed equilibrium

that we see here in the Hawk-Dove Game is typical of biological situations in which

organisms must break the symmetry between two distinct behaviors, when consistently

adopting just one of these behaviors is evolutionarily unsustainable.

We can interpret the result of this example in two possible ways. First, all participants

in the population may actually be mixing over the two possible pure strategies with the

given probability. In this case, all members of the population are genetically the same, but

whenever two of them are matched up to play, any combination of D and H could potentially

be played. We know the empirical frequency with which any pair of strategies will be played, but not what any two animals will actually do. A second interpretation is that the mixture

is taking place at the population level: it could be that 1/3 of the animals are hard-wired

to always play *D*, and 2/3 are hard-wired to always play *H*. In this case, no individual is

actually mixing, but as long as it is not possible to tell in advance which animal will play *D*

and which will play *H*, the interaction of two randomly selected animals results in the same

frequency of outcomes that we see when each animal is actually mixing. Notice also that

in this case, the fitnesses of both kinds of animals are the same, since both *D* and *H* are

best responses to the mixed strategy $p = 1/3$. Thus, these two different interpretation of the

evolutionarily stable mixed strategy lead to the same calculations, and the same observed

behavior in the population.

There are a number of other settings in which this type of mixing between pure strategies has been discussed in biology. A common scenario is that there is an undesirable, fitnesslowering

role in a population of organisms — but if some organisms don't choose to play this

role, then everyone suffers considerably. For example, let's think back to the Virus Game in

Figure 7.2 and suppose (purely hypothetically, for the sake of this example) that the payoff

when both viruses use the strategy #H2 were (0.50, 0.50), as shown in Figure 7.7.

Virus 1

Virus 2

#6 #H2

#6 1.00, 1.00 0.65, 1.99

#H2 1.99, 0.65 0.50, 0.50

Figure 7.7: The Virus Game: Hypothetical payoffs with stronger fitness penalties to #H2.

In this event, rather than having a Prisoner's Dilemma type of payoff structure, we'd have

a Hawk-Dove payoff structure: having both viruses play #H2 is sufficiently bad that one of them needs to play the role of #6. The two pure equilibria of the resulting two-player game

7.6. EXERCISES 225

— viewed as a game among rational players, rather than a biological interaction — would be

(#6,#H2) and (#H2,#6). In a virus population we'd expect to find an evolutionarily stable

mixed strategy in which both kinds of virus behavior were observed.

This example, like the examples from our earlier discussion of the Hawk-Dove Game

in Section 6.6, suggests the delicate boundary that exists between Prisoner's Dilemma and

Hawk-Dove. In both cases, a player can choose to be "helpful" to the other player or "selfish".

In Prisoner's Dilemma, however, the payoff penalties from selfishness are mild enough that

selfishness by both players is the unique equilibrium — while in Hawk-Dove, selfishness is

sufficiently harmful that at least one player should try to avoid it.

There has been research into how this boundary between the two games manifests itself

in other biological settings as well. One example is the implicit game played by female

lions in defending their territory [218, 327]. When two female lions encounter an attacker

on the edge of their territory, each can choose to play the strategy *Confront*, in which she

confronts the attacker, or *Lag*, in which she lags behind and tries to let the other lion confront

the attacker first. If you're one of the lions, and your fellow defender chooses the strategy

Confront, then you get a higher payoff by choosing *Lag*, since you're less likely to get injured.

What's harder to determine in empirical studies is what a lion's best response should be to a

play of *Lag* by her partner. Choosing *Confront* risks injury, but joining your partner in *Lag*

risks a successful assault on the territory by the attacker. Understanding which is the best

response is important for understanding whether this game is more like Prisoner's Dilemma or Hawk-Dove, and what the evolutionary consequences might be for the observed behavior

within a lion population.

In this, as in many examples from evolutionary game theory, it is beyond the power of

current empirical studies to work out detailed fitness values for particular strategies. However,

even in situations where exact payoffs are not known, the evolutionary framework can

provide an illuminating perspective on the interactions between different forms of behavior

in an underlying population, and how these interactions shape the composition of the

population.

7.6 Exercises

1. In the payoff matrix below the rows correspond to player A's strategies and the columns

correspond to player B's strategies. The first entry in each box is player A's payoff and

the second entry is player B's payoff.

Player A

Player B

x y

x 2, 2 0, 0

y 0, 0 1, 1

226 *CHAPTER 7. EVOLUTIONARY GAME THEORY*

(a) Find all pure strategy Nash equilibria.

(b) Find all Evolutionarily Stable strategies. Give a brief explanation for your answer.

(c) Briefly explain how the sets of predicted outcomes relate to each other.

2. In the payoff matrix below the rows correspond to player A's strategies and the columns

correspond to player B's strategies. The first entry in each box is player A's payoff and

the second entry is player B's payoff.

Player A

Player B

x y

x 4, 4 3, 5

y 5, 3 5, 5

(a) Find all pure strategy Nash equilibria.

(b) Find all Evolutionarily Stable strategies. Give a brief explanation for your answer.

(c) Briefly explain how the answers in parts (2a) and (2b) relate to each other.

3. In this problem we will consider the relationship between Nash equilibria and evolutionarily

stable strategies for games with a strictly dominant strategy. First, let's define

what we mean by *strictly dominant*. In a two-player game, strategy, X is said to be a

strictly dominant strategy for a player i if, no matter what strategy the other player j

uses, player i's payoff from using strategy X is strictly greater than his payoff from any

other strategy. Consider the following game in which $a, b, c,$ and d are non-negative

numbers.

Player A

Player B

X Y

X a, a b, c

Y c, b d, d

Suppose that strategy X is a strictly dominant strategy for each player, i.e. $a > c$ and

$b > d$.

(a) Find all of the pure strategy Nash equilibria of this game.

(b) Find all of the evolutionarily stable strategies of this game.

(c) How would your answers to parts (a) and (b) change if we change the assumption

on payoffs to: $a > c$ and $b = d$?

7.6. EXERCISES 227

Player A

Player B

X Y

X 1, 1 2, x

Y x, 2 3, 3

4. Consider following the two-player, symmetric game where x can be 0, 1, or 2.

(a) For each of the possible values of x find all (pure strategy) Nash equilibria and all

evolutionarily stable strategies.

(b) Your answers to part (a) should suggest that the difference between the predictions

of evolutionary stability and Nash equilibrium arises when a Nash equilibrium uses a

weakly dominated strategy. We say that a strategy s'_i is weakly dominated if player i

has another strategy s''_i with the property that:

(a) No matter what the other player does, player i's payoff from s''_i is at least as large

as the payoff from s'_i, and

(b) There is some strategy for the other player so that player i's payoff from s''_i is

strictly greater than the payoff from s'_i.

Now, consider the following claim that makes a connection between evolutionarily

stable strategies and weakly dominated strategies.

Claim: Suppose that in the game below, (*X,X*) is a Nash equilibrium and

that strategy *X* is weakly dominated. Then *X* is not an evolutionarily stable

strategy.

Player A

Player B

X Y

X a, a b, c

Y c, b d, d

Explain why this claim is true. (You do not need to write a formal proof; a careful

explanation is fine.)

Game theory and evolution: ®nite size and absolute ®tness

Keywords: Evolutionary stable strategy; Evolutionary formalism; Evolutionary entropy; Demographic variance;

Reproductive potential; Correlation index

1. Introduction

Evolutionary game theory, as proposed by Maynard Smith and Price [1], is a mathematical

formalism which aims to explain in evolutionary terms how con¯icting interests can lead to stable

behavioral traits within a population of organisms. In this game-theoretic model, individuals in

the population are said to adopt strategies, which represent any genetically programmed behavioral

action. The payo€ associated with a strategy is described by the Darwinian ®tness of the

individuals that adopt it. A strategy is said to be evolutionarily stable if it has the property that if

all members of the population adopt it, no mutant phenotype could invade the population under

the in¯uence of natural selection.

Mathematical Biosciences 168 (2000) 9±38

www.elsevier.com/locate/mbs

* Corresponding author.

E-mail addresses: ldemetr@oeb.harvard.edu (L. Demetrius), gundlach@math.uni-bremen.de (V.M. Gundlach).

PII: S0025-5564(00)00042-0

The notion of an evolutionarily stable strategy (ESS) represents the cornerstone of evolutionary

game theory: the concept has been given mathematical formalizations in several di€erent contexts

(cf. [2±6]) and attempts have been made to characterize its properties and explicitly specify the

strategy in a large class of examples.

The mathematical representation of the model rests on two main assumptions:

A(1) Population size is large ± e€ectively in®nite. This condition ensures that ¯uctuations in

population numbers due, for example, to small perturbations in the individual payo€s can be

neglected ± a situation which ensures that conditions for the invasion of a mutant strategy will be

a deterministic process.

A(2) The ®tnesses ascribed to individual strategies refer to relative values, such as the change in the mean number of o€spring that survive for reproduction, rather than absolute values, such as

the net reproduction rate.

In the analysis of models based on the assumptions of in®nite size and relative ®tness measures,

strategies are parametrized in terms of the distribution p . .pi., where pi represents the frequency

of individuals adopting the ith strategy. The notion of an evolutionarily stable strategy is characterized

in terms of the selective advantage s of the incumbent population with respect to the

mutant type, which is given by

s . DW ; .1:1.

where, W, the index of Darwinian ®tness, is measured in terms of numbers of surviving o€spring.

This notion of selective advantage has been invoked by Maynard Smith [2], Hofbauer and

Sigmund [3], Zeeman [4], Hines [5] and Lessard [6] to show that an ESS satis®es two properties:

B(1) An equilibrium condition ± the Nash solution concept. This refers to the strategy p . .pi., where each player holds the correct expectation about the others behavior and acts to maximize

his ®tness.

B(2) A stability criterion. The incumbent strategy is evolutionarily stable if given that the

population is at a Nash equilibrium, the ®tness of the incumbent always exceeds the ®tness of any

rare mutant.

The class of games considered by Maynard Smith and Price are distinguished by two main

features: (a) they deal with local competition within a population, (b) they pertain to situations in

which the success of a strategy depends on its frequency. The mathematical formalism, in particular

the measure of ®tness invoked, is adapted to analyze an evolutionary dynamics which

refers to interactions of individuals within populations.

As emphasized, however, in [7], a complete evolutionary theory must also address itself to

problems which arise from the interaction of the population with the external environment. The questions which arise in this context are: what is the probability of extinction of a population in

the given environment? Under what conditions will a mutant, that is, a population whose phenotypic

composition di€ers slightly from the incumbent population, invade the existing type and

replace it. Lewontin [7] seems to have been the ®rst to recognize that these problems can be

formalized in a game theory context, which is now called Games against Nature. In this class of

models, the selective unit is a population which is de®ned by a certain behavioral feature or

strategy, and the central problem concerns the invulnerability of the population to invasion by

populations adopting mutant strategies. Games against Nature deals with interactions between

populations rather than interactions within populations. This class of games and the evolutionarily

10 L. Demetrius, V.M. Gundlach / Mathematical Biosciences 168 (2000) 9±38

stable strategies (ESS) they generate have been studied recently by several authors, in particular

Metz, Nisbet and Geritz [8], Rand et al. [9], Geritz [10], Rees and Westoby [11]. The analytical

representation of this class of models revolves around two main conditions:

C(1) Population size is large, e€ectively in®nite ± a condition identical to A(1) and ensuring

similar e€ects.

C(2) The ®tness assigned to strategies refers to absolute values such as net-reproduction rate,

rather than relative values such as the change in the mean number of o€spring that survive for

reproduction.

The selective advantage s, in this class of models is given by

s . Dr; .1:2.

where r represents the rate of increase in total population numbers, given by the Malthusian

parameter in constant environment models, and the dominant Lyapunov exponent in random environment contexts. The derivation of (1.2) can be shown to rest implicitly on the assumption

that the population size of the incumbent is e€ectively in®nite.

This article will relax the in®nite size condition and analyze Games against Nature on the

assumption that size is ®nite. We will appeal to a statistical mechanics formalism, elaborated in

other population dynamics contexts in [12,13], to show that in this new class of models, the selective

advantage s of the incumbent strategy with respect to the mutant strategy will now be

represented in terms of two demographic variables, the population growth rate r, the rate of

increase of total population numbers, and the demographic variance r2, which represents the

variance in the contribution of the di€erent strategists to the total payo€.

Analytically, selective advantage s will be shown to be described by

s . Dr ÿ 1

N

Dr2; .1:3.

where N denotes the total population size. The derivation of (1.3) will be based on a stochastic

model which considers the invasion condition of new mutants in the resident population to be

determined by a di€usion process, on account of the ⁻uctuations due to the ®nite size condition.

We refer to [14], where stochastic elements are introduced in the analysis of the mutation process

in models distinct from the cases considered in this paper.

In our analysis of this new family of models we will parametrize the population in terms of the

distribution I . .li

., where li represents the relative contribution of the ith strategy to the total

payo€. We will denote by M the collection of such distributions and appeal to the new characterization

of selective advantage to show that ESS is this new context satis®es the following analogues

of B(1) and B(2).

D(1) An equlibrium condition, which we call thermodynamic equilibrium, is described in terms

of the function evolutionary entropy H de®ned by

H . ÿ

X

i

li log li: .1:4.

Evolutionary entropy H is a measure of the diversity of options associates with a strategy ± pure

strategies have zero entropy, mixed strategies positive entropy. Thermodynamic equilibria refer to

L. Demetrius, V.M. Gundlach / Mathematical Biosciences 168 (2000) 9±38 11

strategies which are extremal states (minima or maxima) of entropy, within a natural class of

strategies known as demographic equilibria.

D(2) A stability criterion, which will be expressed in terms of two demographic parameters, the

reproductive potential U and the correlation index c. The parameter U measures the net-payo€, averaged over all strategy classes; whereas c is an index of the correlation, with respect to the

di€erent strategy classes, of the net payo€ function.

We will appeal to this depiction of ESS in terms of the equilibrium condition and stability

criteria to establish the following set of criteria for ESS.

Theorem 1.1. Evolutionary stable strategies are characterized by extremal states of entropy.

There exist only ®nitely many strategies for which entropy becomes extremal. If there are no

ecological constraints restricting strategies to subsets Mc of M, the extremal strategies are global

and described by the pure strategies, which minimize entropy, and the equidistribution on X

which maximizes entropy. In the case of constraints, extremal strategies will be local and de®ned

on the boundary of the subsetMc. Hence the evolutionarily stable strategies constitute a subset of

these three classes of strategies: the equidistributed strategy, pure strategies and non-equidistributed

extremal strategies.

In the case of strategies that are global maxima or minima, we will show that the existence

criterion can be expressed in terms of a single parameter, the reproductive potential.

Theorem 1.2. An equidistributed strategy is an ESS if and only if it yields a negative reproductive

potential.

Theorem 1.3. A pure strategy is an ESS if and only if it yields a positive reproductive potential.

In the case of constraints limiting the number of strategies for which entropy becomes extremal,

the excluded global maximum or minima, respectively, can be replaced by local maxima or

minima appearing on the boundary of the set Mc. Then the existence criteria will be expressed in

terms of the two variables, the reproductive potential, and the correlation index.

Theorem 1.4. A non-equidistributed extremal strategy is an ESS if and only if either of the following

conditions holds:

(i) U60; $c > 0$;

(ii) UPO; c < 0:

Our analysis will establish that non-equidistributed extremal strategies are local maxima if and

only if the reproductive potential is negative and the index is positive, whereas, they describe local

minima if and only if the reproductive potential is positive and the index is negative.

This article is organized as follows. In Section 2 we describe in detail the concepts and formalism

we have introduced to analyze the game theoretic models in which the notions of ®nite size

and absolute ®tness measures are considered. In particular, we provide a motivation for the

notion thermodynamic equilibrium. In Section 3 we present a perturbation analysis of such

equilibria which forms the basis for our analysis of the mutation event and our investigations of

12 L. Demetrius, V.M. Gundlach / Mathematical Biosciences 168 (2000) 9±38

ESS. We exploit the notions of Section 2 and the results of Section 3 to derive our main results, namely Theorem 1.1 (Section 4) and Theorems 1.2±1.4 (Section 5). Section 6 contrasts the formalism

of the classical models (in®nite size, relative ®tness) with the new set of models (®nite size,

absolute ®tness). We illustrate the signi®cance of our new formalism by a study of the evolution of

the sex ratio, see [15], and the evolution of polymorphism in seed size, see [10,11]. Studies of the

sex ratio in the context of classical game theory models predict a 1±1 sex ratio is the ESS. Our

analysis based on evolutionary entropy as the measure of ®tness predicts that a 1±1 sex ratio is

evolutionarily stable if and only if the population has stationary size. Our analysis thus indicates

that in exponentially growing populations departures from a 1±1 sex ratio may occur. Studies of

seed size polymorphism [10], in terms of game theory models invoking growth rate as ®tness

predict that there is no stable seed size monomorphism: the polymorphic condition is the unique

ESS. We will appeal to the entropy formalism to generate a larger repertoire of ESS. We will

delineate three classes of ESS, each class being de®ned by the ecological condition the plants

experience. The three classes of ESS are described as follows:

(a) Equal number of seeds of di€erent sizes.

(b) All seeds of the same size.

(c) Variable number of seeds of di€erent sizes.

The polymorphic strategy (a) and the monomorphic strategy (b), both require the stringent

condition that seed size and reproductive yield are uncorrelated. We will show that strategy (a) is

an ESS if the total net-reproductive yield is bounded by the number of distinct seed sizes, whereas

strategy (b) is an ESS if the net seed production increases exponentially. The polymorphic strategy

(c) requires that seed size and reproductive yield are correlated.

We should remark at this point that the notion of entropy has been applied in other population dynamics contexts, see for example the work of Bomze [16], Iwasa [17] and Ginzburg [18]. In these

studies, entropy is interpreted primarily in terms of the Kullback±Lieber distance and the properties

which form a central part of our work were not exploited by these authors. The entropy

concept used in this paper is embedded in ergodic theory and statistical mechanics, a formalism

which was introduced in a population biology context in [19]. The thermodynamic methods we

exploit, invoke besides entropy, a large family of related macroscopic variables ± reproductive

potential, demographic variance. The power of the application of the thermodynamic formalism

to game theory resides in the pertinence of these concept to characterize the complexities generated

by problems of ®nite size and absolute measures of ®tness.

2. Games against Nature

We consider a large but ®nite population of individuals. We assume that the behavior of the individuals in the population can be described in terms of a set of choices X :. fx1; . . . ; xdg. These

elements xi represent the di€erent options which are available to the individuals in the population.

The payoff associated with the set X should depend on the choices of the xi only, in particular it

should be independent of the time the choice is made. Hence it is represented by a function u

which assigns to each xi 2 X a real non-negative number u.xi., a measure of the net-o€spring

production associated with the option xi.

L. Demetrius, V.M. Gundlach / Mathematical Biosciences 168 (2000) 9±38 13

The quantity Z.u. . Pd

i.1 u.xi. represents the total net-o€spring production of a given individual

in the population. Let N.n. denote the number of individuals in the population at time n.

Then

N.n . 1. . Z.u.N.n.:

Hence

N.n . 1. . Z.u.nN.0.:

The net reproduction rate, that is, the growth rate per generation, denoted r.u. is given by

r.u. . lim

n!1

1

n

log N.n.;

hence

r.u. . log Z.u.: .2:1.

2.1. Demographic equilibria

We consider a population of organisms parametrized in terms of behavioral options x1; . . . ; xd

for a game. The net-o€spring production associated with the option xi is u.xi.. From (2.1) we

observe that subject to the process de®ned by u.xi., the population will increase per generation at

the rate r.u..

Now let M denote the set of probability distributions l on X. An element l 2M ascribes to each xi a number li, with 06li 61,

Pd

i.1 li

. 1. The quantity I . .li

. is called a strategy. It

refers to the preference of certain options in the populations and not to individual action; i.e. we

consider the class of strategies I 2 M as the expression of a demographic program which assigns

u.xi. to each behavioral option xi 2 X. Hence a game can be described by the mathematical object

.X; I;u., where I 2M.

We now introduce the notion of a demographic equilibrium by showing that the parameter r.u.

satis®es a variational principle analogous to the principle of the minimization of free energy in

statistical mechanics, see [13,20]. We de®ne for any strategy I 2 M, the evolutionary entropy H

given by (1.4), and the reproductive potential U given by

U :. l.log u. .

Xd

i.1

li logu.xi.: .2:2.

We introduce for any l 2 M, the quantity

P.l;u. . H.l. . l.logu.

and call it the power of the strategy l corresponding to the payo€ u. Then a probability distribution

^l 2 M is said to be a demographic equilibrium state if

P.^l;u. . sup

l2M

P.l;u.:

Thus at demographic equilibrium the strategy maximizes the sum of the entropy and reproductive

potential. We will now show that the maximum P.^l;u. is precisely r.u., the growth rate given by

14 L. Demetrius, V.M. Gundlach / Mathematical Biosciences 168 (2000) 9±38

(2.1). In view of this property, the concept demographic equilibrium characterizes the state of a population which is increasing exponentially at the rate r.u..

By appealing to [20, 0.2], we can formalize these observations and characterize the demographic

equilibrium condition in terms of the following well-known proposition.

Proposition 2.1. A strategy ^l is a demographic equilibrium state corresponding to the payoff u if

one of the following two equivalent conditions is satisfied:

(i) ^l satisfies

r.u. . H.^l. . ^l.logu.; .2:3.

where r.u. is the population growth rate.

(ii) ^l . .^li

. is given by

^li

. u.xi.

Z.u. : .2:4.

Proof. By the de®nitions of power and entropy we have for any strategy l

P.I; u. . ÿ

Xd

i.1

li log li

.

Xd

i.1

li logu.xi. .

Xd

i.1

li log

u.xi.

li

:

In order to make comparisons between u.xi. and li easier let us normalize u by setting ~u.xi. :.

u.xi.=Z.u. such that ~u.xi.61 and

P

i ~u.xi. . 1: Then

P.I; u. .

Xd

i.1

li log

~u.xi.

li

. log Z.u.

and the ®rst term on the right-hand side becomes maximal if and only if the fraction is constant

for all i, i.e. identical to 1 because of the normalization. This yields ^li

. u.xi.=Z.u.. _

Let us summarize that the setup for a game is completely described by the payo€ function u

de®ned on the space of options X. Our model and analysis then shows that a demographic

equilibrium represents a biological meaningful state where the three parameters r; H; U characterize the game. In the evolutionary studies we develop, we will assume that the population is at

demographic equilibrium. The system can therefore be represented by the mathematical object

.X; ^l; u. with the macroscopic parameters de®ned at the state ^l. The relation between the three

parameters r; H; U from (2.3) given by

r . H . U:

We note from the above identity that

$U < 0) r < H; U > 0) r > H:$

Accordingly, we will use the parameter U to classify populations according to prevailing constraints

on their growth rate: $U < 0$ (bounded growth), $U > 0$ (unbounded growth).

L. Demetrius, V.M. Gundlach / Mathematical Biosciences 168 (2000) 9±38 15

2.2. Evolutionarily stable strategies (ESS)

The incumbent population can be described in terms of the triple .X; ^l;u.. In fact, ^l is given by

(2.4) and thus completely determined by u, as are the demographic parameters r.u.; H and U due to Proposition 2.1.

We will consider a mutation to be represented by a change in the distribution of net-o€spring

production. Such a change in distribution, which we denote by u_ will induce a new demographic

equilibrium strategy ^l_. Hence the mutant population can be represented by the triple

.X; ^l_; u_..

Mathematically we think of u_ as a small perturbation of u. Smallness is expressed in terms of a

parameter d of small absolute value such that u_ . u.d.. We postulate that the simplest equation

describing the potential u.d. generated by a mutation in a gene that de®nes a phenotype with

potential u, will assume the form

log u_ . log u.d. . log u . d logu: .2:5.

Expression (2.5) simply asserts that logu.d. is a sum of two components: (a) the ®rst, log u due to

the ancestral type, (b) the second, d log u due to some deviation d of small absolute value from the ancestral type. The natural generalization of (2.5) is

$$\log u.d. . \log u . d \log w; .2:6.$$

that means

$$u_- . uwd; .2:7.$$

where

$$Z$$

$$\log u \, d^\wedge l .$$

$$Z$$

$$\log w \, d^\wedge l .2:8.$$

and

$$d$$

$$dd$$

$$Z$$

$$\log u \, d^\wedge l_-$$

$$--$$

$$d.0$$

$$. d$$

$$dd$$

Z

logw d^l_

— —

d.0

: .2:9.

Condition (2.8) asserts that the rates log u and log w averaged over the di€erent strategies

coincide, while (2.9) asserts that the perturbation mainly changes intensities of the rates. The

generalization (2.6) thus states that the component d logw which characterizes the mutation is

now represented by a deviation d from a type with the same average growth rate as the ancestral

type.

The notion of an evolutionarily stable strategy ^l for the resident population .X; ^l;u. requires

that the population .X; ^l_; u_. described by any deviant strategy ^l_ will be displaced in

competition, by the one described by ^l. This condition can be analytically expressed as follows.

Let N_.n. and N.n. denote the total payo€ up to time n 2 Z. for the mutant and incumbent

strategy, respectively. As we intend to consider in the selection procedure N.n. and N_.n. not only

as deterministic, but also as stochastic processes due to ‾uctuations, we rather work with general

N.n. and N_.n. and not just with N.n. . Z.u.n and N_.n. . Z.uwd.n. Due to the ®nite sizes of

16 L. Demetrius, V.M. Gundlach / Mathematical Biosciences 168 (2000) 9±38

both payo€s, in particular the small size of the mutant payo€, ‾uctuations in their values will

occur. We identify the total payo€s of the incumbent and mutant with their population size.

Hence the frequency p.n. of the mutant is given by

p.n. . N_.n.

N_.n. . N.n. : .2:10.

We assume that the mutant population is initially rare, i.e. N_.n. _ N.n. for small n. Then ^I is an

ESS if p.n. ! 0 as n ! 1for any mutant population.

We should emphasize at this point that this new criterion for an ESS rests critically on the

assumption that population size is ®nite (as a condition for the use of the stochastic model), and

that ®tness is described by absolute population properties. We will appeal to this characterization

of ESS to show that ESS are given by extremal states of entropy.

Example. We will illustrate the concepts we have introduced by a reformulation of the sex-ratio

game. A complete analysis of the game will be described in Section 7.

The set of choices is given by X . $fx1; x2g$, where $x1; x2$ represent the pure strategies of producing

male or female o€spring. Let $N1$ and $N2$ denote the population size in the daughter and

grand daughter generation, respectively. The payo€s $u.x1.$ and $u.x2.$ are given by

$u.x1. . p$

m

$N2$

N1

; u.x2. . 1 ÿ p

1 ÿ m

N2

N1

: .2:11.

Here m denotes the average sex ratio in the population and the ratio p : 1 ÿ p represents the

distribution of male and female o€springs produced by an individual. The total net-o€spring

production is given by Z.u. . u.x1. . u.x2., hence

Z.u. . p

m

N2

N1

. 1 ÿ p

1 ÿ m

N2

N1

: .2:12.

The state l . .l1; l2

. which de®nes demographic equilibrium is given by

l1

. u.x1.

Z.u. ; l2

. u.x2.

Z.u. ;

where u.x1.; u.x2. are given by (2.11) and Z.u. is given by (2.12). The parameters reproductive

potential U and evolutionary entropy are de®ned at demographic equilibrium by

U . p.1 ÿ m.

p . m ÿ 2mp

log

p

m

$N2$

$N1$

$--$

$.\,m.1\,ÿ\,p.$

$p\,.\,m\,ÿ\,2mp$

$\log$

$1\,ÿ\,p$

$1\,ÿ\,m$

$N2$

$N1$

$--$

$;\,.2:13.$

$H\,.\,ÿ\,p.1\,ÿ\,m.$

$p\,.\,m\,ÿ\,2mp$

$\log$

$p.1\,ÿ\,m.$

$p\,.\,m\,ÿ\,2mp$

$ÿ\,m.1\,ÿ\,p.$

p . m ÿ 2mp

log

m.1 ÿ p.

p . m ÿ 2mp

: .2:14.

L. Demetrius, V.M. Gundlach / Mathematical Biosciences 168 (2000) 9±38 17

The state ^l . .^l1; ^l2

. which de®nes what we call the thermodynamic equilibrium is given by an

extremal state of the entropy function, de®ned by (2.14).

We will show that there is a unique extremum ± it corresponds to the state which maximizes H.

Indeed, the analysis of (2.14) will show that the maximum value of H is attained when p . m. This

condition corresponds to the thermodynamic equilibrium,

^l . .1=2; 1=2.: .2:15.

The state de®ned by (2.15) is equidistributed. Accordingly, we will exploit the condition $U < 0$,

where U is de®ned by (2.13), to show that the equidistribution (2.15) is an ESS if and only if

$N2=N1 . 1$, a condition which corresponds to a stationary size constraint.

These observations indicate that the .1

2 ; 1

2

. sex ratio can be realized in terms of a simple optimality

principle, namely the maximization of evolutionary entropy.

3. Perturbation analysis

An incumbent population at demographic equilibrium is described by the mathematical

object $.X; \hat{}l;u.$, where $\hat{}l$ denotes the demographic equilibrium which corresponds to the payo€

u. At demographic equilibrium, the population is described by the parameters $r; H; U$ and r2. The latter is called demographic variance and will be introduced shortly. The deviants

derived by mutation in a small subset of the incumbent population will be assumed to be also

at demographic equilibrium. This population will be represented by the object .X; ^ld;u.d.. for

d of small absolute value, where u.d. is a perturbation of the payo€ u and ^ld the demographic

equilibrium corresponding to the payo€ u.d.. The mutants will be described by

the parameters r.d.; H.d.; U.d. and r2.d.. We will now exploit the perturbation analysis

described in [13], in order to evaluate the change Dr; DH; Dr2 which are induced by the

mutation event.

We consider mutations de®ned by (2.7) satisfying conditions (2.8) and (2.9). We obtain for the

mutation process, for ®xed w satisfying Eqs. (2.8) and (2.9), the new function

~r.d. :. r.uwd. . log Z.uwd. . log

Xd

i.1

elogu.xi.

ed logw.xi.: .3:1.

This function is clearly analytic in d and we have

~r0.d. . 1

Z.uwd.

d

dd

Z.uwd.

. 1

Z.uwd.

Xd

i.1

u.xi.w.xi.d logw.xi.

.

Xd

i.1

log w.xi.^ld

.xi.;

18 L. Demetrius, V.M. Gundlach / Mathematical
Biosciences 168 (2000) 9±38

~r00.d. .

Xd

i.1

log w.xi. d

dd

^ld

.xi.

. ÿ 1

Z.uwd.2

d

dd

Z.uwd.

_ _2

. 1

Z.uwd.

$d2$

$dd2\ Z.uwd.$

$.\ ÿ\ 1$

$Z.uwd.2$

Xd

$i.1$

$u.xi.w.xi.d\ logw.xi.$

Xd

$k.1$

$u.xk.w.xk.d\ log\ w.xk.$

$.\ 1$

$Z.uwd.$

Xd

$i.1$

$u.xi.w.xi.d.logw.xi..2$

$.\ ÿ$

Xd

$i.1$

log w.xi.^ld

.xi.

!2

.

Xd

i.1

.log w.xi..2^ld

.xi.;

~r000.d. . 2

Z.uwd.3

d

dd

Z.uwd.

_ _3

ÿ 3

1

Z.uwd.2

d

dd

Z.uwd.

– –

d2

dd2 Z.uwd.

– –

. 1

Z.uwd.

d3

dd3 Z.uwd.

. 2

Xd

i.1

logw.xi.^ld

.xi.

!3

ÿ 3

Xd

i.1

log w.xi.^ld

.xi.

Xd

k.1

.log w.xk..2^ld

.xk.

.

Xd

i.1

.log w.xi..3^ld

.xi.;

in particular

~r0.0. . 1

Z.u.

Xd

i.1

u.xi. logw.xi. .

$$\int \log w \, d\hat{l}; \quad .3:2.$$

where $\hat{l}$ is the demographic equilibrium state for u. We note that when $u . w$, then we have

$$r0.0. .$$

$$\int \log u \, d\hat{l}: \quad .3:3.$$

Furthermore,

$$\sim r00.0. .$$

$$\int .\log w.2 \, d\hat{l} \quad \ddot{y}$$

$$\int \log w \, d\hat{l}$$

$$\_ \_2$$

$$.$$

$$\int \log w$$

$$\_$$

ÿ

Z

logw d^l

_2

d^l .: r2.w.P0: .3:4.

Analogously, it follows that

~r000.0. . 2

Z

log w d^l

_ _3

ÿ 3

Z

logw.2 d^l

Z

logw d^l .

Z

.logw.3 d^l

.

Z

logw

–

ÿ

Z

logw d^l

_3

d^l .: j.w.: .3:5.

L. Demetrius, V.M. Gundlach / Mathematical Biosciences 168 (2000) 9±38 19

Note that from the variational principle (2.3) and (3.1) we can deduce that ~r.d. .

H.ld

. .ld

.log.uwd.., i.e.

~r.d. . H.ld

. . ld

.log u. . dld

.logw.

and hence

~r0.0. . H0.^l. . d

dd

ld

.log u.j

d.0

. ^l.log w.: .3:6.

Comparing (3.2) and (3.6) we obtain

H0.^l. . ÿ d

dd

ld

.log u.j

d.0

and consequently

o2

oeod

r.u1.ewd.j

e.d.0

. d

dd

ld

.logu.j

d.0

. ÿH0.^l.: .3:7.

Let us remark that r2.w., as de®ned in (3.4), is the variance of the random variable log w with

respect to the equilibrium state ^l for u.

We note that

r2.u. .

Z

.log u.2 d^l ÿ

Z

log u d^l

_ _2

.3:8.

and we have r2.u. . 0 () H . 0 or H . Hmax, where Hmax denotes the maximal entropy given

by log d and realized by the equidistribution (see for example [21, Corollary 4.2.1]). We call r2.u.

the demographic variance of the strategy l, and for ®xed u we denote by r2.d. the demographic

variance of the equilibrium state ld for u1.d. Note that

r2.d. . .1 . d.2

Z

.log u.2 d^ld

"

ÿ

Z

log u d^ld

_ _2

#

:

So, if we choose w . u, then

r2.d. . .1 . d.2~r00.d.

and therefore we obtain via di€erentiation with the help of (3.5) that

dr2

dd

.0. . 2r2.u. . j.u. .: c: .3:9.

Moreover we deduce from (3.7) that

r2 :. r2.0. . r00.0. . HO.^l.: .3:10.

We call c given by (3.9) the correlation index of the strategy. The parameter c is a measure of the

correlation with respect to the di€erent strategy classes of the net-payo€ function. This quantity c

is 0 for both pure strategies and equidistributed mixed strategies.

So, for d of small absolute value we have the following approximations by the linearizations

(3.3), (3.9) and (3.10):

20 L. Demetrius, V.M. Gundlach / Mathematical Biosciences 168 (2000) 9±38

Dr _ Ud; DH _ ÿr2d; Dr2 _ cd:

In particular, we obtain for d of small absolute value the following so-called mutation relations:

$U < 0$) DrDH > 0; .3:11.

$U > 0$) DrDH < 0; .3:12.

$c < 0$) DHDr2 > 0; .3:13.

$c > 0$) DHDr2 < 0: .3:14.

Let us remark that in contrast to r2, which is always non-negative, c can assume negative and

positive values.

We will illustrate the values assumed by H; r2 and c with an example which will play an important

role in the analysis of the sex-ratio model, namely any two-option game with strategies

I . .p; 1 ÿ p. for p 2 .0; 1. and normalized payo€s, i.e. u . l. It can easily be checked that the

normalization u.xi. 7!u.xi.=.u.x1. . u.x2..; i . 1; 2 can be done without loss of generality, as

this procedure does not change the quantities of interest.

Any way, in this (and thus in the general two-option game) case we have

H . ÿp log p ÿ .1 ÿ p. log.1 ÿ p.;

Fig. 1. Entropy for two-option games.

L. Demetrius, V.M. Gundlach / Mathematical Biosciences 168 (2000) 9±38 21

r2 . .1 ÿ p.p log

p

1 ÿ p

_ _2

;

c . .1 ÿ p.p log

p

1 ÿ p

_ _2

2

_

. .1 ÿ 2p. log

p

1 ÿ p

—

:

Fig. 1 shows the population entropy for a two-option game as a function of p, while Fig. 2

presents the demographic variance and Fig. 3 the evolutionary index as a function of p.

4. Evolutionarily stable strategies: characterization

In this section we derive necessary and su□cient conditions for the existence of an ESS. Our

analysis revolves around the assumption that population sizes are ®nite and the idea that owing to

statistical ⁻uctuations the quantities $N.n.$ and $N\_.n.$ described in Section 2 are random variables.

Consequently, invasion of the mutant strategy will be a stochastic event.

Our treatment of this stochastic analysis invokes three elements:

1. An investigation of $N.n.$ on the microscopic level by studying sample paths and stochastic

processes;

Fig. 2. Demographic variance for two-option games.

22 L. Demetrius, V.M. Gundlach / Mathematical Biosciences 168 (2000) 9±38

2. The introduction of a stochastic model for N.n. induced by ̄uctuations caused by the stochastic

behavior on the microscopic level and the approximation of the probability density function

of N.n. in terms of solutions of di€usion equations.

3. A study of the invasion±extinction dynamics of the mutant strategy in terms of a pair of coupled

Fokker±Planck equations.

The e€ect of statistical ̄uctuations due to ®nite population size has an extensive literature which

goes back to the pioneering studies of Wright and Fisher, see for example [22,23] and the review

article by Feller [24]. For other developments which invoke the Fokker±Planck equation we refer

to [25±27].

Statistical ̄uctuations in population numbers have two main sources:

1. Demographic stochasticity. Sampling e€ects due to the condition of ®nite size. This randomness

arises because of chance realizations of individual probabilities of birth and death. Since independent

individual events tend to average out in large populations, the e€ects of demographic

stochasticity decreases with increasing population size.

2. Environmental stochasticity. Randomness in the case is ascribable to temporal environmental

changes. The magnitude of the population ‾uctuations induced is set by the degree of environmental

variance, independent of the absolute population size.

Each of these mechanisms engender di€erent kinds of variance of the ‾uctuations induced. Hence,

although the Fokker±Planck equations which emerge from the models describing the di€erent

Fig. 3. Correlation index for two-option games.

L. Demetrius, V.M. Gundlach / Mathematical Biosciences 168 (2000) 9±38 23

mechanisms are formally similar, the parameter spaces in which they operate are unique to the model system investigated.

In the studies we undertake, the stochasticity derives from sampling e€ects due to ®nite size.

Thus our description can be subsumed under demographic stochasticity. The variance which this

stochasticity generates will be given by what we call the demographic variance as expressed by

(3.4).

Fluctuations induced by demographic stochasticity is analogous to the internal variations

observed in certain physical processes and our analysis in Section 4.1 draws from the ideas reviewed

in [28]. The development in Sections 4.2 and 4.3 has analogues in the work of Feller [24]

and Gillespie [25] who analyzed related models.

Feller's work gives a general review of di€usion theory in genetics and studies in particular a

two-dimensional branching process as a genetic model. Our formulation in Section 4.2, in

particular, is inspired by this work. Gillespie [25] exploited Feller's ideas to analyze the dynamics of selection in ®nite populations. The expressions we derive in Section 4.3 have their

analogues in [25]. The mathematical models, however, are quite distinct. Our models are game

theoretic systems with individuals adopting mixed strategies. The variance in these systems

corresponds to demographic variance, as de®ned by (3.8); growth rate refers to the asymptotic

rate of increase of the total payo€ as de®ned by (2.1). The models described in [25] are Wright±

Fisher models in population genetics. The variance parameter refers to the variance in the

number of o€spring of a given genotype; the growth rate parameter pertains to the mean

number of o€spring.

4.1. Microscopic analysis of payo€ and growth rate

Let us recall the de®nition of the total payo€ up to time n as

N.n. . Z.u.n .

Xd

ii;...;in.1

u.xi1

.u.xi2

. _ _ _u.xin

.:

Thus N.n. consists of all possible payo€s for options played in n consecutive games. Since there

exist dn di€erent options for such n games, it might become intractable to describe them

completely. As in statistical mechanics we can give up this microscopic description and work

with macroscopic quantities which can be obtained by a probabilistic description of the single

payo€s. The natural probabilistic description of our situation is guaranteed by the use of the

equilibrium strategy I. The nth product of I de®nes a probability measure on Xn equipped with

its power-set as r-algebra and the nth product of u becomes a random variable on that

probability space.

Now let us note that with respect to I the function log u also de®nes a random variable with

®nite state space and with mean U and variance r2 according to the last section. So, let us consider

the nth partial sum of independent and identically distributed (namely according to l) random

variables Yi :. logu and denote it by Sn. This de®nes a sample path Sn : Xn ! R of length n for

logu, the time mean .n ! 1. of which tends to U by the ergodic theorem for almost all x 2 XN

with respect to the product measure P of l on XN. To the time ﬂuctuations .1=n.Sn ÿ U we can

apply the central limit theorem to obtain that

lim

n!1

P

Sn ÿ nU

r

⍰⍰⍰

n

p

—

6 x

—

.N.x.; .4:1.

where N denotes the standard normal distribution. If we de®ne a family of stochastic processes

by

Xn.t. . S.nt. ÿ .nt.U

n

p ;

where ._. denotes the Gau_ bracket, then it is known (see for example [29, 7.3]) that Xn converges

in distribution as n ! 1to Brownian motion X with variance r2.

Thus we can conclude that asymptotically the deviations of the sample path Sn from the mean

for n ! 1for P ± almost all x 2 XN can be approximated by Brownian motion with variance r2t,

if we consider a continuous time evolution St, t 2 R..
In the following we will adopt this continuous

time approximation.

4.2. The di€usion equation

We have just seen that our deterministic model is
related to certain stochastic processes on a

microscopic level. Due to our assumptions that
population sizes are ®nite and ®tness is described

by absolute measures, ¯uctuations should also be
present at the macroscopic level. Therefore, it is

more realistic to replace the deterministic model
given by N.n. . Z.u.n by a stochastic one, which

in fact is not a di€usion approximation, but makes
use of the results of our approximations and

yields the deterministic model for in®nite population
size. Though it might be a bit confusing, we

will denote the new stochastic process by N.t. and
the old deterministic system by N.n..

On the basis of the last section we could think of N.n.
. Z.u.n as a quantity built out of

random variables (according to the equilibrium state
^l) asymptotically ¯uctuating around a mean

value approximately like a Brownian particle. We assume that these internal ﬂuctuations on the

microscopic level cause ﬂuctuations of the same stochastic nature on the macroscopic level thus

transforming the total payoﬀ in a stochastic process. Since we made a continuous time approximation

of the ﬂuctuations, we could also describe the time evolution of N.n. into a continuous

process. As the time-one mapping of the diﬀerential equation

dN

dt

. r.u.N; N.0. . 1

yields N.n. . Z.u.n and since we assume that the ﬂuctuations should have mean 0, we obtain for

the resulting stochastic process N.t.

lim

Dt!0

1

Dt

EfN.t . Dt. ÿ N.t. jN.t. . Ng . r.u.N;

where we denote by E the expectation induced by I. This gives the ®rst in®nitesimal moment of the

process. In order to derive the second in®nitesimal moment of the process we have to specify the

intensity of the variance of the ¯uctuations. As the in¯uence of microscopical ¯uctuation on the

macroscopic level decreases with the size N.t., we assume according to the central limit result of

the last section that the change in log N.t. in the time interval Dt due to ¯uctuations is caused by

Brownian motion with variance r2.u.Dt=N.t. being a non-trivial function of time. This form of

L. Demetrius, V.M. Gundlach / Mathematical Biosciences 168 (2000) 9±38 25

the variance also guarantees that for in®nite population size the process N.t. is just the continuous

time version of N.n.. Thus we obtain for the second in®nitesimal moment of the process

lim

Dt!0

1

Dt

Ef.N.t . Dt. ÿ N.t..2 jN.t. . Ng . N2 lim

Dt!0

1

Dt

r2Dt

N

. r2N:

Consequently we can deduce from [30, II] that the solution of the Fokker±Planck equation

of

ot

. ÿr

o.fN.

oN

. r2

2

o2.fN.

oN2

yields the density f .N; t. for the process N.t., which can be viewed as the solution of the stochastic

di€erential equation

dN . rN dt . r

???

N

p

dWt :

Note that we could also describe the evolution of N_.n. corresponding to the mutant population

in an analogous way. Indeed, the density f _.N_; t. for the process N_.t. becomes

of _

ot

. ÿr o.f _N_.

oN_

. r_2

2

o2.f _N_.

oN_2 :

It will be our goal to compare the stochastic processes N.t. and N_.t., in particular we want to

analyze the process p.t. . N_.t.=.N.t. . N_.t... The result will in general be di€erent from the

one derived from a deterministic approach, where the limit of p.t. as t ! 1can be completely

determined in terms of Dr and hence U only. In the deterministic approach, the selective advantage

is evidently given by

s . Dr:

As we have argued before, such a deterministic consideration would have to be based on in®nite

population size (for t ! 1) and would not take care of ¯uctuations, which should be present. In

the stochastic representation, size is ®nite and the selective advantage will be shown to be determined

by

$s \cdot Dr \ddot{y} 1$

M

$Dr2;$

where $M \cdot N \cdot N\_$, the total population size.

We note that when M tends to 1, the selective advantage for the stochastic representation

coincides with that for the deterministic representation. Moreover, our stochastic process would

coincide for discrete time with the original deterministic system, if the population size were in-

®nite. Therefore we propose that the more suitable process to consider for evolutionary phenomena

is the stochastic one, which is not a di€usion approximation of the original system

anymore, but which presents in our view an improved model for the evolutionary dynamics of

interest.

4.3. The invasion condition

We are looking now at the stochastic model just developed and consider the evolutionary dynamics of the mutant and incumbent populations when the mutant is rare. During this phase,

26 L. Demetrius, V.M. Gundlach / Mathematical Biosciences 168 (2000) 9±38

the numbers N and N_ of the incumbent and mutant population are assumed to be separately

regulated. Let w.N; N_; t. denote the bivariate density of the pair .N.t.; N_.t... Invoking the

statistical independence condition, it follows that the density w is given by w.N;N_; t. .

f .N; t.f _.N_; t., where f _ is the density function corresponding to the mutant population. Since

f .N; t. and f _.N_; t. are solutions of Fokker±Planck equations, we have that w satis®es the equation

ow

ot

. ÿr

o.wN.

oN

. r2

2

o2.wN.

oN2

ÿ r_ o.wN_.

oN_

. r_2

2

o2.wN_.

oN_2 :

Recalling (2.10) and de®ning M.t. . N.t. . N_.t., we obtain

ow

ot

. ÿo.a.p;M.w.

op

. 1

2

o2.b.p;M.w.

op2

ÿ o.f.p;M.w.

oM

. 1

2

o2.d.p;M.w.

oM2

. o2.X.p;M.w.

op oM

;

where

a.p;M. . p.1 ÿ p. Dr

–

ÿ 1

M

Dr2

–

; .4:2.

b.p;M. . p.1 ÿ p.

M

.r2p . r_2.1 ÿ p..; .4:3.

f.p;M. . M.pr_ . .1 ÿ p.r.;

d.p;M. . M.pr_2 . .1 ÿ p.r2.;

X.p;M. . p.1 ÿ p.Dr2:

We are mainly interested in the problem of invasion or extinction of the mutant population. This

decision is taken for small N_ _ M and N close to equilibrium. Consequently, the changes in N_

and N will be very small and mainly at the expense of the other. This implies that the changes inM

will be negligibly small, in particular in e€ecting the result of the invasion±extinction problem.

Thus we restrict our attention to the case where we can neglect the derivatives of w with respect to

M. We set a.p;M. _ a.p.; b.p;M. _ b.p., and now consider

ow

ot

. ÿo.a.p.w.

op

. 1

2

o2.b.p.w.

op2 : .4:4.

The problem of extinction±®xation can be analyzed by considering the Kolmogorov backward

equation

ow

ot

. a.p. ow

ot

. 1

2

b.p. o2w

op2

with the boundary conditions

w.0; t. . 1; w.1; t. . 0:

Let P.y. denote the ultimate probability that the mutant becomes extinct, when y denotes the

initial frequency of the mutant, i.e. P.y. . limt!1 w.y; t.. Then P.y. satis®es the ordinary differential

equation

L. Demetrius, V.M. Gundlach / Mathematical Biosciences 168 (2000) 9±38 27

a.y. dP

dy

. 1

2

b.y. d2P

dy2

. 0 .4:5.

with boundary conditions P.0. . 1, P.1. . 0. Write

G.x. . exp

—

ÿ 2

Z x

0

a.y.

b.y. dy

_

:

Then, from (4.5), we obtain

P.y. .

R 1

y G.x.dx

R 1

0 G.x.dx

:

Write

s . Dr ÿ 1

M

Dr2: .4:6.

Then using the expressions for a.p. and b.p. given in (4.2) and (4.3), G.x. becomes

G.x. . 1

—

ÿ Dr2

r_2 x

_2Ms=Dr2

and hence

P.y. . 1

—

ÿ Dr2

r_2

_.2Ms=Dr2..1

ÿ 1

—

ÿ Dr2

r_2 y

_.2Ms=Dr2..1

,

1

—

ÿ Dr2

r_2

_.2Ms=Dr2..1

ÿ 1: .4:7.

In order to determine the shape of this function, we take a look at the derivatives

P0.y. . Dr2

r_2

2Ms

Dr2

—

. 1

—

1

—

ÿ Dr2

r_2 y

_2Ms=Dr2,

1

—

ÿ Dr2

r_2

_.2Ms=Dr2..1

ÿ 1;

P00.y. . ÿ Dr2

r_2

_ _2 2Ms

Dr2

—

. 1

—

2Ms

Dr2 1

—

ÿ Dr2

r_2 y

_.2Ms=Dr2.ÿ1

,

1

—

ÿ Dr2

r_2

_.2Ms=Dr2..1

ÿ 1:

Except in the degenerate ± and for large M unusual ± case of 2Ms=Dr2 . ÿ1, P0.y. cannot vanish,

hence must be negative. In this situation the numerator of P00.y. is always negative, while the

denominator is negative for $s > 0$ and positive for $s < 0$. Thus we can conclude the following:

$s > 0$) P is convex;

$s < 0$) P is concave:

The degree of curvature of P depends on the magnitude of s; i.e. on the values of Dr; Dr2 and M,

as can be seen from the following graphs, obtained by numerical calculation for some suitable

choices of Dr2=r_2 and Ms=Dr2 and showing the dependence of P.y., the ultimate probability that

the mutant becomes extinct, on y, the initial frequency of the mutant.

28 L. Demetrius, V.M. Gundlach / Mathematical Biosciences 168 (2000) 9±38

If DrDr2 < 0, then the sign of s does not depend on M and a reasonable large M leads to an

exponent b :. .2Ms=Dr2. . 1 . .2MDr=Dr2. ÿ 1 su⊠ciently small to cause the extreme shape

shown in Fig. 4. If Dr > 0; Dr2 > 0, we have to distinguish two cases: for Dr > Dr2=M we have

s > 0 with exponent b su⊠ciently large for reasonable M leading to the extreme shape shown in

Fig. 5(a), while for Dr < Dr2=M we have s < 0 with an exponent that increases in M and hence

leads to a probability which is an increasing function of M (cf. Fig. 5(b)). If Dr < 0; Dr2 < 0 (cf.

Fig. 6), we have to distinguish two analogous cases leading to the following assertions about the

extinction±invasion phenomenon for populations of a size M:

1. $Dr > 0$; $Dr2 < 0$: the mutant invades almost surely.

2. $Dr < 0$; $Dr2 > 0$: the mutant becomes extinct almost surely.

3. $Dr > 0$; $Dr2 > 0$:

· $M > Dr2 = Dr \_ c = U$: the mutant invades almost surely,

· $M < Dr2 = Dr \_ c = U$: the mutant becomes extinct with a probability which decreases as M

increases.

4. $Dr < 0$; $Dr2 < 0$:

· $M > Dr2 = Dr \_ c = U$: the mutant becomes extinct almost surely,

· $M < Dr2 = Dr \_ c = U$: the mutant invades with a probability which increases as M increases.

We can deduce from (4.6) that for ®nite population size M invasion can be excluded if

$Dr < 0$; $Dr2 P0$ or $Dr60$; $Dr2 > 0$ .4:8.

with respect to any mutant strategy. We will in Section 4.4, exploit the mutation relations given in

Section 3 to show that the criteria (4.8) can be characterized in terms of the extremal states of

entropy.

Fig. 4. (a) $Dr > 0$; $Dr2 < 0$ and (b) $Dr < 0$; $Dr2 > 0$.

L. Demetrius, V.M. Gundlach / Mathematical Biosciences 168 (2000) 9±38 29

4.4. ESS as extremal states of entropy

We recall that a strategy is an ESS if it is invulnerable to invasion by any deviant strategy. We

have seen that such an ESS is guaranteed under condition (4.8) with respect to any mutant

strategy. We can specify such situations as follows.

Fig. 5. $Dr > 0$; $Dr2 > 0$ and (a) $M > Dr2=Dr$, (b) $M < Dr2=Dr$.

Fig. 6. $Dr < 0$; $Dr2 < 0$ and (a) $M > Dr2=Dr$, (b) $M < Dr2=Dr$.

30 L. Demetrius, V.M. Gundlach / Mathematical Biosciences 168 (2000) 9±38

Proposition 4.1. Relations (4.8) can hold only for any deviant strategy if and only if one of the

following two cases occurs:

(i) $U < 0$; cP0 or U60; $c > 0$;

(ii) $U > 0$; c60 or UP0; $c < 0$:

In the first case (4.8) is equivalent to DH 60; in the second to DH P0.

Proof. We recall from (3.11) and (3.12) that $U < 0$ implies DrDH > 0; and $U > 0$ implies

DrDH < 0. We also recall from (3.13) and (3.14) that c < 0 entails DHDr2 > 0 and c > 0 implies

DHDr2 < 0. Hence in the case $U < 0$ we have that Dr < 0 implies DH 60. In addition, when

cP0, we have that Dr2P0 implies DH 60. Analogously it follows that in the case of $U > 0$ we

have Dr < 0 entails DH P0; and when c < 0, we observe that Dr2P0 implies DH P0. We also

get the respective assertions if we start with the strict inequalities for c. Hence when (i) holds, (4.8)

is equivalent to DH 60, and when (ii) holds, (4.8) is equivalent to DH P0. _

This proposition provides not only a nice characterization of ESS, but also a necessary condition

for them. Since the conditions DH 60 and DH P0 assert that the incumbent strategies

correspond to maximum and minimum states of entropy, respectively, we immediately obtain

from Proposition 4.1 the following result.

Theorem 4.2. ESS are described by extremal states of entropy.

Hence every ESS is an extremal state of entropy. However not all extremal states are ESS. In

order to delineate the class of extremal states which are ESS, we need to examine the stability

criteria as described by the parameters U and c.

5. Evolutionarily stable strategies: existence

We have managed so far to derive a characterization of possible ESS by appealing to entropy.

We will now make use of this characterization to obtain an existence result, i.e. conditions

for the existence of an evolutionary stable strategy. In deriving these existence criteria we note that evolutionary entropy H is bounded from below and from above, and strategies,

where H attains its maximum and minima are well characterized (see for example [21,

Corollary 4.2.1]).

Lemma 5.1. For a set X of d choices every game (X; l; u) satisfies H.l.6 log d; and

(i) H.l. . log d if and only if li

. 1=d for all i . 1; . . . ; d; i.e. l is the equidistribution,

(ii) H.l. . 0 if and only if li

. 0 for all i . 1; . . . ; d except one j with lj

. 1; i.e. l is a pure

strategy.

This lemma yields a pre-selection for ESS. Due to constraints some of these candidates for an

ESS cannot be realized and have to be replaced by points on the boundary of the set Mc of

possible strategies. Thus we can immediately deduce the following result.

L. Demetrius, V.M. Gundlach / Mathematical Biosciences 168 (2000) 9±38 31

Proposition 5.2. If there are no constraints on the payoffs the only possible candidates for an ESS

are the pure strategies and the equidistribution. In case of constraints excluding any of these candidates,

these have to be replaced by strategies which maximize or, respectively, minimize entropy

and which necessarily have to lie on the boundary of the set Mc of admissible strategies.

We know that the evolutionary stable strategies form a subset of the three extremal strategies:

the unique global maximum, the global minima, and the local extrema. This third situation arises

when certain constraints on the payo€ function obtain. Our existence result can now be expressed

in terms of the following theorems.

Theorem 5.3. The equidistribution on X is an ESS if and only if it satisfies $U < 0$.

The equidistribution constitutes the global maximum. The condition $U < 0$ simply expresses a stability condition. The equidistribution satis®es c . 0 ± a property which implies that there exists

no correlation between the strategy classes and the net-payo€ function.

Theorem 5.4. A pure strategy is an ESS if and only if it satisfies $U > 0$.

Pure strategies represent global minima: the relation $U > 0$ expresses a stability criterion. Pure

strategies also satisfy the condition c . 0, which, as we observed, de®nes the case where the

strategy classes and net-payo€ function are uncorrelated.

Theorem 5.5. In the case of constraints, a local maximum can replace the excluded global maximum

if and only if

$U < 0; c > 0;$

and a local minimum can replace an excluded global minimum if and only if

$U > 0; c < 0;$

Non-equidistributed mixed strategies can represent local maxima or minima: since these extremal states are local, the index c which describes the stability property may have arbitrary

sign. In the case of a local maximum, the condition c < 0 describes a negative correlation between

the strategy classes and the net-payo€ function, the condition c > 0 represents a positive

correlation.

At ®rst sight, the result of Theorem 4.2 might not look very interesting, as it yields rather

strange looking ESS. This is due to the formal character of our approach, in particular the general

setup of our games. In general, our ESS contain implicitly more conventional information about

the evolutionarily stable strategy, in particular if there exist constraints for the payo€s, a situation

described by the condition c 6. 0. We are going to illustrate this in the analysis of two well-studied

games, see Section 7.

6. Games within populations, Games against Nature: a contrast

It is of some interest at this point to contrast the classical models based on games within

populations with the new class of models based on Games against Nature developed in this paper.

6.1. Games within populations: in®nite size, relative ®tness

The models proposed by Maynard Smith and Price [1] are concerned with pairwise con¯icts

within a population. The game can be described in terms of a set of options X . fx1; . . . ; xdg. The

state of the population is described by the vector p . .pi., where pi is the frequency of i-strategists.

If an i-strategist contends with a j-strategist in a pairwise con¯ict, the payo€ to the former is aij,

where aij is measured in terms of an increase in ®tness (the expected number of surviving o€-

spring). An i-strategist will encounter a j-strategist with frequency pj and then receive a payo€ aij.

The expected payo€ to the i-strategist is

Pd

j.1 aijpj. The expected payo€ to a population p . .pi.

against itself is

$$E.p; p. .$$

$$X$$

$$i;j$$

$$piaijpj:$$

The expected payo€ to population q . .qi. against p . .pi. is

$$E.q; p. .$$

$$X$$

$$i;j$$

$$qiaijpj:$$

A state p is said to be an ESS if, when all individuals in the population adopt this strategy, no

mutant can invade. To characterize this analytically, we consider a population consisting mainly

of individuals playing strategy p, with a small frequency e of some mutants playing an alternative

strategy p_. The ®tness of the population of incumbents W .p. is given by

$$W .p. . .1 ÿ e.E.p; p. . eE.p; p_.:$$

The ®tness of the population of mutants is

W .p_. . .1 ÿ e.E.p_; p. . eE.p_; p_.:

Selective advantage s is given by s . DW , where DW . W .p_. ÿ W .p..

The condition that p is an ESS requires s < 0. This property holds if the following conditions

are satis®ed:

(i) E.p; p.PE.p_; p. for all p_, that is

X

i;j

p_

i aijpj 6

X

i;j

piaijpj; .6:1.

an equilibrium condition ± the Nash criterion.

(ii) If E.p; p. . E.p_; p. and p_ 6. p, then E.p_; p_. < E.p_; p., that is

X

i;j

p_

i aijp_

j <

X

i;j

piaijp_

j ; .6:2.

a stability condition.

L. Demetrius, V.M. Gundlach / Mathematical Biosciences 168 (2000) 9±38 33

6.2. Games against Nature: ®nite size, absolute ®tness

The models considered in this article are concerned with games described by payo€s to certain

behavioral options which could be interpreted as interactions between the population and an

external environment. Thus, the population is parametrized in terms of a set of d options and characterized by a corresponding distribution representing the net-o€spring production.

A strategy is the probability distribution ^l given by

^l . u.xi.

Z.u. ; Z.u. .

Xd

i.1

u.xi.:

The ®tness variables in this class of models are the population growth rate r, the demographic

variance r2, and the entropy H. The incumbent population in these models is described by the

mathematical object .X; ^l; u.. A mutant population is represented by the object .X; ^l_; u_., where

u_ is given by

log u_ . log u . d log u

and l_ by

^l_

i

. u_.xi.

Z.u_. :

The selective advantage s in this model is now expressed by (4.6). The condition that ^I is an ESS

requires that s < 0 for all M, where M denotes the population size of the incumbent. This

property entails that

Dr < 0; Dr2P0 or Dr60; Dr2 > 0:

These conditions hold if the following relations are satis®ed:

1. I is an extremal state of entropy H ± an equilibrium condition.

2. I is evolutionary stable ± a stability condition. Stability is expressed in terms of conditions on

the reproductive potential U and the index c. When I is a global maximum, we require U < 0; a

global minimum U > 0. When I is a local maximum, we require U < 0; c > 0; a local minimum,

then U > 0; c < 0.

The contrast between the two classes of models is summarized by the Table 1.

Table 1

Relation between the classical and non-classical models

Game theory

properties

Games within populations

(in®nite size, relative ®tness measures)

Games against Nature

(®nite size, absolute ®tness measures)

Measures of ®tness Mean payo€ E.p; p. . Pd

i;j.1 piaijpj Entropy H

Conditions of ESS Either E.p; p. > E.p_; p. or

E.p; p. . E.p_; p.; E.p_; p. > E.p_; p_.

Either Dr < 0; Dr2P0 or Dr60; Dr2 > 0

Solution concept Nash equilibria E.p; p.PE.p_; p.

for all p_

Thermodynamic equilibria: extremal states

of H

Stability condition If E.p; p. . E.p_; p.; p_ 6. p, then

E.p_; p_. < E.p_; p.

DH 60, U < 0, cP0, DH P0, U > 0, c60

34 L. Demetrius, V.M. Gundlach / Mathematical Biosciences 168 (2000) 9±38

7. Examples

We illustrate the signi®cance of our new formalism by studies of the evolution of the sex ratio

and the evolution of seed size polymorphism, two phenomena that have been studied in the

context of classical models of evolutionary game theory.

7.1. Sex-ratio model

In this section we reformulate a classical model, the sex-ratio game, in order to illustrate the

applicability of our existence and characterization theorems. We will show that the .1

2 ; 1

2

. sex ratio

observed in many populations can be characterized in terms of an optimality principle, the

maximization of evolutionary entropy. Our analysis will show that an ESS exists if and only if the

population has a stationary size. The stationary condition which emerges from our study can thus

explain departures from .1

2 ; 1

2

. sex ratios observed in many natural populations. Such departures

have often been attributed to genetic or behavioral constraints, see for example [15]. Our model

indicates that ecological factors which permit rapid exponential growth can also create anomalous

sex ratios.

Let us consider a sexual population of one species and parametrize it by the state space

X . fM; F g. In our context M and F represent the pure strategies of producing just male or just female o€spring, respectively. We will assume as in [3] that generations are non-overlapping. We

also assume that the population is described by a polymorphic group of individuals. Let m denote

the average sex ratio in the population, N1 the population size in the daughter generation, and N2

the population size in the granddaughter generation. We will assume N2PN1 to exclude the trivial

case where the population becomes extinct. As shown in [3, 15.4], the expected number of children

produced by a male in the ®rst generation is N2=mN1, the expected number of children produced

by a female in that generation is N2=.1 ÿ m.N1. Let us assume that the primary sex ratio is determined

by genes acting on the homogametic sex. As pointed out by Hamilton [31], a measure of

the propagation of the gene will now be measured by the expected number of grandchildren. An

individual in the population will produce male and female o€spring according to his strategy in

the ratio p : 1 ÿ p. The payo€ function is thus given by (2.11). The expected number of grandchildren

Z.u. is given by (2.12) and the thermodynamically stable strategy $\hat{I}$ by

$\hat{I}_1$

. p.1 ÿ m.

p . m ÿ 2mp

; $\hat{I}_2$

. m.1 ÿ p.

p . m ÿ 2mp

;

leading to the reproductive potential given by (2.13) and the evolutionary entropy given by (2.14).

Note that p is the primary sex ratio which is a genetically determined quantity, while m represents

a secondary sex ratio, which is observed in the population and depends on cultural and

environmental factors.

Let us start our investigations of evolutionary stable strategies for the sex-ratio model with the search for candidates on the basis of our characterization via evolutionary entropy. Since the pure

strategies correspond only to trivial, i.e. degenerate cases like p . 0; p . 1; m . 0 or m . 1, we

will restrict our attention to the only non-trivial possible case given by the maximally mixed

strategy. In this case, evolutionary entropy becomes maximal for the equilibrium ^l if and only if

L. Demetrius, V.M. Gundlach / Mathematical Biosciences 168 (2000) 9±38 35

^l.M. . ^l.F. . 1=2. Note that the distribution ^l . .1

2 ; 1

2

. represents the demographic strategy.

The ratio p : 1 ÿ p which represents the proportion of male and female o€spring produced satis®es

p

1 ÿ p

. m

1 ÿ m

:

Since the function f .x. . x=.1 ÿ x. is strictly increasing on .0; 1., we can in fact derive that an

ESS for the sex-ratio model has to satisfy p . m. On the basis of this partial result, we can now

analyze the existence of an ESS. Namely, for p . m the reproductive potential is

U . log

N2

N1

and hence non-positive if and only if N2=N1 61. Since we assumed N2=N1P1, we have the

following result.

Corollary 7.1. The unique non-trivial ESS for the sex-ratio model is obtained for p . m; if and only

if N2=N1 . 1.

Corollary 7.1 has signi®cant implications for human demography. We predict that during episodes of `baby boom', a condition de®ned by a large positive growth rate, signi®cant departures

from the .1

2 ; 1

2

. sex ratio should occur. The only case we know of where the relation between population

growth and sex ratio has been documented is [32]. This study of the Israel population during

the period 1943±1953, a period of rapid exponential growth, is consistent with our predictions.

7.2. Seed size polymorphisms

Game theoretic models, based on growth rate as a measure of ®tness, have been proposed to

explain the enormous variation in seed size which describe plant species. Geritz [10] for example,

has shown that a single seed size is never evolutionary stable: there is always selection for some

continuous variation in seed size. We will apply the game theoretic models based on evolutionary entropy as ®tness to show that there exist three classes of seed size distributions that are evolutionarily

stable. Our analysis will show that (i) when correlations between seed size and net-reproductive

yield vanishes, there exist two evolutionarily stable seed size patterns: (a) equal number

of seeds of di€erent size ± when population growth is bounded by the number of seed sizes; (b) all

seeds of the same size ± when total seed production increases; (ii) when a correlation between seed

size and net-reproductive yield is obtained, there exists a unique evolutionarily stable seed size

pattern, namely a continuous variation in seed size.

Let X denote the set of seed sizes, X . fx1; . . . ; xdg. Let u.xi. denote the net-reproductive yield,

that is net production associated with seed size xi. The quantity Z.u. .

P

i u.xi. represents the

total net-seed production of an individual. Population growth rate r.u. is given by r.u. . log Z.u.. The strategy ^l . .^li

. that de®nes demographic equilibrium is given by

^li

. u.xi.=Z.u., where ^li represents the proportion of seeds of sizes xi produced by an individual.

The evolutionary stable strategies are extremal states of entropy H . ÿ

P

i ^li log ^li. To characterize

these strategies, we consider the reproductive potential U .

P

i ^li log u.xi.. This quantity

measures the net-seed production averaged over all the size classes. We observe that

36 L. Demetrius, V.M. Gundlach / Mathematical Biosciences 168 (2000) 9±38

U . r ÿ H:

Hence U < 0 entails r < H and describes a bounded population growth; also U > 0 implies r > H and represents unbounded population growth. The expression

r2 .

Xd

1

li

.logu.xi..2 ÿ

Xd

1

li log u.xi.

!2

is the variance in the net-reproductive yield. The quantity c . 2r2 . j, where j is given by (3.5)

represents the correlation between the seed size and net-reproductive yield. The ESS are characterized

by evaluating the extremal states of entropy. We now characterize these extremal states

and we appeal to Theorem 5.3±5.5 to determine whether they correspond to ESS.

(i) The distribution I . .1=d; . . . ; 1=d.: equal number of seeds of different sizes. This is the global

maximum of H. For this distribution we have c . 0 and H . log d. The total net reproduction is

Z.u.. Since log Z.u. . H . U, the reproductive potential U is given by

U . log

Z.u.

d

:

From Theorem 5.3 we infer that the equidistribution is an ESS if and only if log.Z.u.=d. < 0, that

is, if and only if Z.u. < d. We conclude that the equidistribution, equal number of seeds of

di€erent sizes, is an ESS, if the following constraints prevail:

(a) Seed size and individual seed production are uncorrelated (c . 0).

(b) Total net-seed production is bounded by the number of seed sizes .Z.u. < d..

(ii) The distribution l . .0; . . . ; 0; 1; 0; . . . ; 0.: all seeds of the same size. This condition describes

a global minimum of H. For this distribution we also have c . 0, and H . 0. The total net reproduction

is Z.u.. In view of the identity log Z.u. . H . U, we have U . log Z.u.. From

Theorem 5.4 we infer that the pure strategy l . .0; . . . ; 0; 1; 0; . . . ; 0. is an ESS if and only if

log Z.u. > 0, that is, if and only if Z.u. > 1. We conclude that the pure strategy de®ned by the

property, all seeds of the same size, is an ESS if the following constraints obtain:

(a) Seed size and individual seed production are uncorrelated (c . 0).

(b) Total net-seed production is greater than unity .Z.u. > 1..

(iii) The distribution l . .li

. with li

6. 1 for all i and li

6. lj for some i; j: a continuous variation

in seed size. This condition describes a local extremum of H. In this case c 6. 0, and 0 < H < log d.

From Theorem 5.5, we infer that the above distribution de®nes an ESS if one of the following

situations obtain:

(a) A negative correlation between seed size and net-reproductive yield .c < 0.; unbounded

growth (U > 0).

(b) A positive correlation between seed size and net-reproductive yield .c > 0.; bounded growth

.U < 0..

www.ingramcontent.com/pod-product-compliance
Lightning Source LLC
Chambersburg PA
CBHW071759200526
45167CB00017B/522